"With great courage and meticulous research, this is a searing indictment of Western moral failure in the face of genocide. De Vogli's unflinching analysis forces us to confront how, by reserving empathy for some but denying it to others, democracies have become complicit in mass atrocity. Essential, urgent, and profoundly necessary, this is a masterwork of extraordinary intellectual rigor and profound moral courage."

    – Richard Wilkinson, Emeritus Professor of Social Epidemiology, University of Nottingham, and author of *The Spirit Level*

• • •

"Mature democracies have entered a dystopia: the smaller the injustice, the more extreme the response; the greater the injustice, the deeper the bureaucratic indifference. De Vogli shows something that compounds the problem: one can easily strip people of empathy by manipulating their consciousness into believing that the victims belong to a different species."

    – Nassim Nicholas Taleb, author of *The Black Swan*

• • •

"A brilliant, powerful, and essential book. It presents a moral challenge we are obliged to confront."

    – George Monbiot, columnist for *The Guardian* and author of *The Invisible Doctrine: The Secret History of Neoliberalism* and *Bring on The Apocalypse: Six Arguments for Global Justice*

• • •

"Selective grievability and white unseeing are dissected in this thought-provoking book on how the genocide was sustained for over two years."

    – Ghassan Abu-Sittah, Rector of the University of Glasgow, Director of the Conflict Medicine Program, Global Health Institute, American University of Beirut

• • •

"The original proposed publisher of this book provided a graphic illustration of its argument, when they refused to allow Roberto De Vogli to even use the word 'genocide'. Since this is the only term which adequately captures what Israel has done to the Palestinians, we must be grateful that he refused and

found a more principled publisher. *Selective Empathy* provides a passionate, incisive and scrupulously documented explanation of the West's complicity, and deserves to be widely read."

– Martin Shaw, Emeritus Professor of International Relations, Institute of International Studies, Barcelona, and author of *The New Age of Genocide* and *What Is Genocide?*

• • •

"In this extremely well-researched and balanced volume, Roberto De Vogli provides updated and systematic knowledge on the horror of the Israeli genocide in Gaza – the killing, the starving, the dehumanizing, the scholasticide, the ecocide. The selective empathy and ensuing denial of solidarity with Palestinians among Western elites is linked to the powerful propaganda of Israeli and Western allies. At the same time, however, courageous acts by activists, doctors, and journalists testify to the emergence of universal empathy and global identification."

– Donatella della Porta, Professor and Founding Dean of the Faculty of Political and Social Sciences, Scuola Normale Superiore, Florence, and author of *Social Movements: An Introduction*

• • •

"Selective empathy is not our destiny is the ethico-political cry that emanates from Roberto De Vogli's powerful and uncompromising book. Whereas Western political powers preach a (neo)colonial ethics of ideological resemblance – care only about Israeli Jewish lives, lives that look like yours – so as to obscure the horrors of the Gaza genocide, De Vogli carefully attends to the ways this message is thwarted by incredible acts of self-sacrifice (from doctors volunteering in Gaza, Israeli soldiers refusing to serve in an eliminationist war, university students protesting on their campuses, and faculty risking their careers for the cause of Palestinian liberation). In *Selective Empathy*, Gaza registers as a site of utter destruction and global resilience. Western leaders desperately want us to forget about the former and stop the latter. De Vogli courageously points us in the opposite direction."

– Zahi Zalloua, Professor of Literature and Philosophy, Whitman College, and author of *Solidarity and the Palestinian Cause: Indigeneity, Blackness, and the Promise of Universality*

• • •

"This riveting must-read offers a heart-rending but rigorous analysis of the systematic dehumanisation of the Palestinian people in Gaza and beyond, the complicity of the international community and major Western powers – and a startling insight into the systemic, institutional and ultimately civilisational processes and values that have led us to a point where genocide becomes fully normalised. I strongly recommend it."
   – Nafeez Ahmed, investigative journalist, *The Guardian* blogger, and author of *Alt Reich* and *Failing States, Collapsing Systems*

• • •

The research underpinning *Selective Empathy* gave rise to the open letter and public appeal "Stop the Silence: Academic Associations Must Recognize the Genocide in Gaza," as well as to an article published in *The Lancet* with Jonathan Montomoli, Ghassan Abu-Sittah and Ilan Pappé, titled "Break the Selective Silence on the Genocide in Gaza." The letter, which gathered more than 15,000 signatures, including 5,500 from university professors, researchers, physicians, and professionals, prompted numerous academic and professional associations, representing millions of members worldwide, to officially recognize the genocide in Gaza.

• • •

"Thank you for organizing this important petition. I am very happy to sign it."
   – Avi Shlaim, Emeritus Professor of International Relations, University of Oxford, and author of *The Iron Wall: Israel and the Arab World*

• • •

"It's an excellent letter!"
   – Rashid Khalidi, Edward Said Professor of Modern Arab Studies, Columbia University, and author of *The Hundred Years' War on Palestine*

• • •

Among the signatories of the open letter:
– Norman Finkelstein, political scientist and author of *The Holocaust Industry*
– Ilan Pappé, historian and author of *The Ethnic Cleansing of Palestine*
– Omer Bartov, Samuel Pisar Chair of Holocaust and Genocide Studies, Brown University, and author of *Anatomy of a Genocide*

- Tomaso Montanari, Rector, University for Foreigners of Siena
- Nadera Shalhoub-Kevorkian, Professor of Global Law, Queen Mary University of London; visiting scholar at Harvard and Princeton
- Yanis Varoufakis, economist, leader of MeRA25 and co-founder of DiEM25; Professor of Economics, University of Athens; Honorary Professor of Political Economy, University of Sydney

Selective Empathy

Studies in Critical Social Sciences

The titles published in this series are listed at *brill.com/scss*

Selective Empathy

The West through the Gaze of Gaza

By

Roberto De Vogli

BRILL

LEIDEN | BOSTON

Originally published in hardback in 2026.

Typeface for the Latin, Greek, and Cyrillic scripts: "Brill". See and download: brill.com/brill-typeface.

ISSN 1573-4234
ISBN 978-90-04-76630-3 (paperback, 2026)
ISBN 978-90-04-74827-9 (hardback)
ISBN 978-90-04-74828-6 (e-book)
DOI 10.1163/9789004748286

This book is printed on acid-free paper and produced in a sustainable manner.

Contents

Preface

June 3, 2024. Imagine receiving this message from the publisher with whom you have already signed a contract to write a book about Gaza: "Dear Roberto, we are delighted that you have accepted our contract proposal for your book 'Selective Empathy.' However, we anticipate problems with the word 'genocide.' Because it is a legal term, with a strict legal definition, the use of this term carries a legal burden and has legal consequences for both you and us as publishers. To avoid this dangerous territory, we would like to ask you to replace 'genocide' with a term that has no legal definition, both for the tagline and for the main text of your book. A possible example would be 'mass atrocities'. Is that okay? Cheers, (signature)."

Of course, it was not okay. But it meant that I no longer had a publisher.

Months later, at a conference in Istanbul, I read the message to a large audience of academics, researchers, and activists. When the session ended, a tall, bearded man wearing red glasses approached me. He was my future editor. After briefly introducing himself, he smiled and said, "We're going to publish your book with the word genocide."

Why a book on selective empathy? And why Gaza? There are so many other wars in the world ...

Selective empathy is a surprisingly underexplored topic in academic literature, despite its profound relevance to understanding psychology, global politics, and humanitarian assistance. The manifestation of this moral and emotional double standard is nowhere more glaring than in the international community's response to Gaza. For two long years, Palestinians have endured extraordinary levels of violence with virtually no international protection or relief. The disparity with the widespread support, aid, and military assistance received by Ukraine is staggering. Each life counts beyond how many can be counted, yet an estimated 17,121 children died in Gaza between October 7, 2023, and July 9, 2025, as a result of Israel's assault (UNRWA, 2025). This is in comparison to the 521 children killed during the Russian invasion of Ukraine between February 24, 2022, and December 31, 2024 (OHCHR, 2025). These numbers do not diminish one tragedy in favor of another, but they do demand reflection on why empathy, and action, are extended so unevenly.

Although some Western politicians and academics have tried to deny the deadly impact of Israel's assault on Gaza, the world's most respected human rights organizations, such as *Amnesty International* and *Human Rights Watch*, have no doubt: Gaza has suffered the crime of all crimes. A genocide. As I write, there have been more than 68,000 deaths (mostly civilians) and more than 170,000 confirmed injuries (UNOCHA, 2025). An analysis published in the *British Medical Journal* comparing child mortality in recent urban conflicts found that "while civilian casualties were significant in all cases studied, none has seen the number of civilian deaths, particularly among children, as in Gaza since last October" (Bhutta et al., 2024). Denying their genocide is like killing them a second time.

Why the West?

While the international community is supposed to be representative of the entire world, Western powers are the most consequential and influential actors in world geopolitics and economics. They influence the direction of world affairs and the decisions of international organizations and humanitarian aid agencies more than any other region in the world. In addition, media pundits and politicians often portray Western civilization as a champion of noble ideals and high values of justice, democracy, and humanity. Yet, these principles have been profoundly betrayed in practice. Moreover, the West is not just a bystander to the genocide in Gaza, but an active military, economic and political player. This makes every Western citizen complicit. While we should care and feel compassion for all victims of any war, our actions and attention should be focused primarily on the war crimes of our governments.

Most of the Western mainstream media described the Gaza assault as an Israeli–Palestinian "conflict" that began on October 7, 2023, when Israel faced what was described as the darkest day in its history. It was undoubtedly a horrific attack, as more than a thousand Hamas fighters and Palestinian gunmen broke through the barriers separating Israel from Gaza and violently assaulted both military targets and civilians. These crimes resulted in approximately 1,200 deaths (including 33 children), 5,400 wounded, and the abduction of 280 civilians and soldiers (UNOCHA, 2025). These appalling atrocities, widely covered by the Western mainstream media, constitute crimes against humanity and have been rightly condemned by the international community and human rights organizations.

Israel then launched a devastating military operation and multiple blockades. The bombardment was systematic and relentless. The blockades deprived Gaza of food, water, fuel, aid, basic medical supplies, and other necessities, resulting in widespread food insecurity, hunger, infectious diseases, and untold human suffering. The destruction of hospitals, bakeries, water and sanitation facilities resulted in a death toll even higher than that caused by direct violence. In front of such an unprecedented human catastrophe, the Western mainstream media have reported on the victims in Gaza very differently from how they have covered the victims of the war in Ukraine, often minimizing, rationalizing, or even justifying the atrocities.

There are so many foreign aggressors in the world. Why Israel?

Because Israel has broken most records of horror in recent history. The brutal assault on Gaza has produced a catastrophic public health crisis. In just one year, the average life expectancy at birth has collapsed by around 35 years, from 75.5 to just 40.5 (Guillot et al., 2025). This represents the most severe decline ever documented within a single year in recent times, even more severe than the drop in life expectancy during the 1994 Rwandan genocide, where there has been a reduction in longevity of 30 years (De Vogli et al., 2025).

As Philippe Lazzarini, Commissioner-General of UNRWA, noted, in the first four months of the prolonged assault on Gaza, Israel's military actions resulted in the deaths of more children than those killed by all other nations at war worldwide in the past four years combined. Israel is responsible for "the highest number of child amputees per capita anywhere in the world" (UNICEF USA, 2025), the destruction of the highest number of hospitals, ambulances, and schools in recent theaters of war worldwide. Israel has massacred the highest number of doctors and medical personnel in conflict zones and killed the highest number of UN workers and humanitarian volunteers in war areas. Gaza is also the largest cemetery of journalists killed in war in the world.

In spite of these horrors, the tragedy in Palestine has not elicited the widespread expressions of global solidarity from Western governments. Why has the West remained indifferent to the most filmed genocide in history? What explains the banality of its complicity?

Noam Chomsky, in what he called the "Orwellian problem," tried to explain the paradox of "knowing so little, even when there is so much evidence," pointing to the role of media propaganda in "manufacturing consent" about geopolitical events and war victims (Chomsky, 1986, p. xxvii). But understanding why Gaza has received so little help and solidarity from the West requires further effort. It is not enough to ask, "Why do we know so little when we have so much evidence?" but also, "Why do we feel so little when we know so much?"

The West's moral inaction and detachment from the suffering of the Gazans cannot be explained simply by a lack of knowledge. What requires explanation is not ignorance, but the absence of compassion and the stunning lack of collective outrage in the face of genocide. At first glance, the West's emotional indifference seems to stem from a failure of empathy. But the West does not respond in this way to all victims of violence. Only some are met with indifference, suspicion, or even justification when they suffer. This is not a lack of empathy per se. It is the result of a self-serving, tribal, and parochial form of emotional response that reserves compassion for some victims while denying it to others.

Selective empathy refers to the tendency to feel compassion primarily or exclusively for members of one's "in-group," based on race, nationality, religion, or social status, while showing little or no empathy for members of the "out-group." This psychological trait is characterized by bias, prejudice, an "us versus them" mentality, and a strong sense of cultural, national exceptionalism.

Of course, the lack of empathy for certain groups is not unique to Western nations. Selective empathy crosses cultural and political boundaries. And some even believe that the West is showing far too much compassion in the world. Elon Musk, for example, has said that "the fundamental weakness of Western civilization is empathy." This claim may be bitterly ironic in regions of the world that have borne the historic brunt of Western invasion, colonization, and exploitation. Many, in the so-called global South, would perhaps resonate with a quote commonly attributed to Gandhi. When asked what he thought of Western civilization, he replied, "I think it would be a good idea."

As Joseph Henrich observed, in his analysis of what he defined as *Western, Educated, Industrialized, Rich and Democratic* (WEIRD) societies, the West often views the rest of the world through a lens tinted by cultural narcissism and a deep-seated sense of superiority (Henrich, 2020a). While not shared by a considerable minority of Western citizens, this mindset frequently reduces other cultures to objects of pity, condescension, or inferiority. It also implies less humanization and emotional consideration. The way Palestinians have been left to die and starve has exposed this psychosocial dynamic with unprecedented clarity. After Gaza, the acronym WEIRD could now stand for *Western, Educated, Indifferent, Rich, and Desensitized.*

The humanitarian crisis in Gaza has exposed the West's crisis of humanity. As we all watched the Palestinian genocide live-streamed, Gaza, in turn, looked back. And what it revealed is profoundly unedifying. In a sense, the image of the West reflected in Gaza's eyes resembles *The Picture of Dorian Gray* by Oscar Wilde: a portrait of moral decay, constantly changing, which the protagonist saw depicted in a hidden painting. In Gaza's gaze, too, the West glimpsed the collapse of its own ethical principles.

But Gaza has also become a litmus test for the world's conscience. One cannot remain neutral in the face of genocide, just as one cannot remain idle on a moving train. There is no middle ground; you are either against genocide or for it. Silence is not neutrality. Silence is complicity. While most Western politicians, mainstream media spokespeople and influential intellectuals have remained indifferent to the pain of Palestine, others have stood up for it. They have been outraged by the horrors suffered in Gaza and have responded with protests, sit-ins, strikes, boycotts, messages, articles, petitions, videos, mass mobilizations. They have forcefully expressed their dissent. Some have paid the price for their solidarity through expulsions, revoked visas, annulled degrees, arrests, and imprisonment. Others even with their lives.

These acts of dissent, which, as the poet Rasha Abdulhadi observes, throw "sand into the gears of genocide," are not only conscientious gestures by citizens who identify with humanity as a whole, but also proof that selective empathy is not inevitable. While nationalism and cultural superiority have characterized the international community's response to the plight of Gaza, some in the West have demonstrated a sense of compassion that transcends national, ethnic, and in-group boundaries.

From the volunteer doctors who risked their lives in what was left of Gaza's last hospitals to the conscientious objectors who refused to join the Israeli army, from the students who have set up encampments for Palestine to the academics who have written, protested, and struck against the genocide, some have stood up to the Western-backed Israeli war machine. While they have not stopped the genocide, they have shown that there is another West that truly cares about equity, solidarity, and global justice.

Of course, selective empathy is in all of us, everyone, everywhere. It is part of our DNA. But selective empathy is not only rooted in our biology. It is also shaped by ideology. If history, culture, education, and socialization have colonized our emotions, then selective empathy is neither our destiny nor an inevitable feeling of our unchangeable nature. It lies mostly in our mental constructions, collective imagination, and manufactured social norms. These can be challenged and changed so that they, in turn, can challenge and change who we are and what we feel.

To empathize and feel the world's pain across all national boundaries sounds utopian. The instinctive response to the emotional indifference of others is often to withdraw our own compassion in return, reinforcing cycles of alienation and deepening the "us versus them" mentality. This is understandable, but selective empathy cannot be fought with more selective empathy. What can we do, then? Will we ever be able to see others as if they belonged to no other country, because the only country is the world? Will we ever be able to perceive humanity as a single entity?

Yes, we can.

Let us take an example. When considering the most innocent victims who have lost their lives in Gaza and Israel since October 7, some Western politicians have called for a few minutes of silence in memory of the Israeli children. Some protesters have rebuffed these calls by highlighting the far greater plight of the Palestinian boys and girls who died in Gaza after October 7. Both missed the point: humanity is not a tribal dimension of our privatized love, but a cosmopolitan, egalitarian idea.

It is dutiful, deeply human, to observe 35 minutes of silence for the 33 Israeli children killed on October 7 and the 2 Israeli children kidnapped and killed as hostages. We must also respect almost 12 days of silence for the 17,121 Palestinian children killed in Gaza.

There is one story that, more than any other, illustrates the difference between selective empathy and universal love.

Imagine a group of children praying in what remains of a destroyed mosque, a mosque with shattered walls and scattered rubble.

Then think of Dr. Mark Perlmutter, a volunteer surgeon from North Carolina, who has tended to the wounded under relentless bombardment in one of Gaza's last functioning hospitals. Imagine him approaching the ghost mosque.

Seeing the children all praying together, Dr. Perlmutter becomes curious and asks the Imam, "What are these children praying for?"

"They are praying for the forgiveness of Israel," replies the Imam.

They were praying to Allah to forgive Israel for what it has done.

Introduction

"We are all Ukrainians" (Camut & Boonen, 2023). These four words reverberated around the world as Russia launched its deadly invasion on February 24, 2022. In those days, the yellow and blue flag became more than a national symbol. It turned into a manifestation of solidarity with the victims of a foreign aggression. For some, humanity seemed to have reached a turning point. Never before had the international community shown such a powerful, coordinated action and sense of compassion for the plight of war victims.

The outpouring of aid, the mobilization of charities, the media attention and the impassioned speeches by government spokespeople showed how the world can react swiftly and decisively to help a nation attacked by a stronger enemy. It also showed how to hold violators of international law to account. The swift sanctions and the boycott of Russian assets were among the drastic measures taken by the international community to defend Ukraine's right to self-determination. These interventions served as a poignant example of how nations, driven by a collective sense of justice, can join forces and influence the course of geopolitical events.

The concept of a robust international community committed to preventing war and promoting peace has deep historical roots. In his 1795 essay *Towards Perpetual Peace*, Immanuel Kant set out a vision for the development of human morality and the ability of nations to secure a stable, lasting peace (Kant, 2006). According to the German philosopher, cosmopolitan justice could only be achieved if all nations agreed to be governed by universal laws, renounced hostility toward other countries, and abandoned their desire for dominance in favor of mutual trust and cooperation.

One might wonder whether the coordination and unity demonstrated by the West on behalf of Ukraine represented a leap forward in the evolution of our moral conscience, in line with the ideals envisaged by Kant. Skeptics, however, refrained from similar naive expectations. They noticed too well and long ago that the international community had often failed to act on behalf of the victims of many past conflicts. Geopolitical interests too often eclipsed solidarity with the victims of some wars.

The invasion of Iraq, for example, which began on March 19, 2003, did not result in a boycott and punishment of the perpetrators of the illegal act. Much like Ukraine, Iraq is a sovereign nation that suffered aggression from a powerful state, yet there were no sanctions imposed on Washington as there were on Moscow. Moreover, the conflict in Iraq has not been met with the same outrage from the international community. Even the response to war refugees has often been very different. Since Russia's 2022 invasion of Ukraine, European states have welcomed millions of Ukrainians while extending far less empathy and protection to people fleeing Afghanistan and Syria (Ajana et al., 2024).

If the use of double standards in expressing solidarity with the victims of some past wars cast doubt on the universality of the values the West claims to cherish, the lack of decisive action to defend civilians during Israel's recent assault on Gaza further exposed the West's collective hypocrisy. But the consequences of the genocide in Gaza go beyond Gaza. At a time when humanity faces unprecedented existential threats, such as the risk of nuclear war and the climate crisis, the genocide has further divided the world. While unprecedented efforts of transnational cooperation and understanding of the perspective of "the other" are urgently needed, impunity for the massacres in Palestine has pushed "the West" further away from "the Rest." The selective humanity that allowed Israel to commit genocide with total impunity and utter contempt for international law has undermined trust in the global community and the principles of human cooperation upon which it is based. Hopefully not beyond repair.

Now, more than ever, the world needs a humanitarian revolution to overcome petty national and geopolitical rivalries based on emotionally destructive drives and ethnocentric worldviews. As civilization stands in the midst of one of the greatest tragedies of our time, we can imagine a different future, one that abolishes war and genocide. This will require profound political, social, and economic transformations, as well as changes in the way we see, feel, and perceive each other.

It is up to us.

Never before has cooperation been so essential to our evolution. The choice before us is dark and stark: either we defend humanity by fostering greater harmony among people and nations, adopting new paradigms of cooperation that transcend nationalism and ethnocentrism, or there will be no humanity to defend.

Prologue

This book offers valuable insight for anyone seeking to understand the West's selective empathy, particularly in its response to the unfolding genocide in Gaza. But its relevance extends beyond this event. It offers a critical and timely exploration of selective empathy itself: how it operates, what drives it, and why it matters.

Adopting an evidence-based, transdisciplinary approach, the book weaves together insights from psychology, geopolitics, media studies, and international relations to examine a pressing yet often overlooked issue. It asks why, in the face of overwhelming evidence of mass suffering, some in the West have taken to the streets in protest and civil disobedience, while others have remained indifferent, complicit, or even supportive of the Gaza genocide. This work offers not only an urgent diagnosis of a moral crisis, but a deeper inquiry into the emotional and cognitive forces that shape how, and for whom, we feel, and care.

The book consists of eight chapters.

Chapter 1 *The Gaze of Gaza* examines the Palestinian humanitarian catastrophe following Israel's prolonged assault after October 7, 2023, with a focus on the staggering civilian death toll, particularly among children. It explores how Western governments and mainstream media have minimized, justified, or denied the genocidal scale of violence, revealing a pattern of selective empathy and geopolitical hypocrisy. The chapter argues that the genocide in Gaza has exposed the moral collapse of Western civilization and its selective application of human rights and dignity.

The second chapter, entitled *Selectively Empathic*, explores the idea of selective empathy, the tendency to empathize only with in-groups while ignoring others. Drawing on psychological and neuroscientific research, it describes the different dimensions of empathy: cognitive, emotional, and motivational. Two schools of thought are examined: one views empathy as a moral necessity, the other sees it as a flawed idea prone to bias. The chapter explains that selective empathy is influenced by both biological and socio-cultural factors. It ends with a discussion of egalitarian, universal empathy rooted in justice and shared humanity.

The third chapter *Why We Fail to Feel* analyzes how entire populations can become selectively indifferent to the suffering of others. It explores the psychopathological mechanisms behind the dehumanization of Palestinians in Israel, and how these foster widespread justification for genocidal violence. The chapter analyzes the abysmal cruelty of the Israeli army, the openly genocidal rhetoric of government officials, and the broad public support for these actions. It further considers how Israeli nationalism weaponized Holocaust memory to sustain oppression. Finally, it critiques Western governments, exposing the banality of their complicity in mass atrocities.

The fourth chapter, entitled *The West and the Rest*, examines the geopolitical and cultural relationship between Israel and Western democracies within a historical context of imperialism and settler colonialism. The chapter explores how Europe and North America came to dominate global political, economic, and military affairs, as well as how cultural perspectives have promoted a Western-centric view of the world. The chapter calls for emotional decolonization to overcome the West's superiority complex. It urges a shift in consciousness that transcends nationalistic bias and selective empathy.

Chapter 5, *Manufacturing Exclusive Compassion*, analyzes the key role of Western mainstream media in shaping people's beliefs and emotions, fostering selective compassion and systemic bias that prioritizes "worthy victims" aligned with Western interests and identity. The chapter discusses evidence showing that disinformation and euphemistic language minimized Israeli war crimes while exaggerating or even fabricating Palestinian ones. A stark media double standard is revealed in comparisons with coverage of Ukraine, where victims were humanized and perpetrators clearly identified.

Chapter 6, *The Herd of Free Thinkers*, focuses on intellectuals who have failed to speak out against the atrocities in Gaza and who refused to recognize these crimes as genocide. The chapter contrasts this hesitance with the consensus among genocide scholars and human rights organizations that have affirmed unequivocally that a genocide took place. The chapter also examines the restrictions on freedom of speech and the various forms of intimidation faced by those who express solidarity with Palestine. Additionally, it examines the conflation of real and false accusations of anti-Semitism, often used to silence criticism and obscure legitimate grievances and realities on the ground.

Chapter 7, *The World's Conscience*, explores conscience as a moral force driving resistance to the Gaza genocide. The chapter begins by highlighting the remarkable heroism of volunteer doctors who have risked or lost their lives to provide care amid overwhelming destruction and danger. The chapter also recounts remarkable acts of dissent, protest, and solidarity. It discusses student protests and grassroots movements that refused to remain silent in the face of genocide. It also documents state violence, censorship, and repression, targeting those showing solidarity with Palestine. The chapter affirms that standing up for justice may not guarantee victory, but it preserves our shared humanity.

The book concludes (Chapter 8 *On the Brink*) with an analysis of how the crisis of selective humanitarianism revealed in Gaza reflects poorly on humanity's urgent need to tackle global threats such as the risk of nuclear war and the climate crisis. The Gaza genocide further divided the West and the Rest at a time when global cooperation, cosmopolitanism, and solidarity are not just essential, but existential. Drawing on evolutionary biology and moral philosophy, the chapter explores whether humans are doomed by tribal instincts or are capable of transcendent empathy. It ends by calling for a collective epiphany, and transformations of conscience and consciousness, a radical shift in identity and values grounded in universal compassion and solidarity.

Some may perceive this book as biased for devoting limited attention to the perspectives of the Israeli government and military. However, these positions have frequently consisted of rationalizations and denials, arguments long recognized as typical of states implicated in mass atrocities. While acknowledging multiple perspectives has value, facts and evidence must prevail. In the case of Gaza, they are unequivocal. In such a scenario, appeals to "both sides" do not serve the cause of fairness and balance. They become a charade; a hollow ritual dressed up as impartiality. When it begins to snow, we do not convene a debate between someone who asserts it is snowing and someone who denies it. We simply look out the window.

Would you invite Slobodan Milosevic to sit next to Srebrenica survivors on a talk show about the Bosnian genocide? Would you place Myanmar military officials across Rohingya refugees to discuss ethnic cleansing? Why then, when it comes to Gaza, do some still insist on giving equal weight to the perspectives of those who suffered genocide and those who perpetrated it?

Why did I write this book?

I felt I had to.

The genocide in Gaza is not just about Palestine and Israel. It concerns the entire world. It is the greatest moral failure of our time. In the aftermath of October 7, when it appeared clear that a genocide was unfolding in Gaza, I felt almost as if I had been living in a state of tragic hypnosis, permeated by a subtle sense of perpetual mourning.

While my taxes were contributing to massacres of innocent people a few thousand miles from where I live, my government kept sending weapons to Israel without my consent. The Western mainstream media denied, trivialized, and rationalized the suffering of Palestinian children killed in Gaza, while my university stayed silent for far too long about the daily atrocities.

I had to write it loud and clear: not in my name.

This book is dedicated to the defenseless women, men, mothers and fathers, patients, the elderly, workers, journalists, volunteers, United Nations (UN) personnel, schoolteachers, university students and professors who have lost their lives in Gaza.

It is also written for the medical doctors, nurses, healthcare professionals, and ambulance drivers who sacrificed their own lives to treat and care for the wounded under the relentless daily bombing of medical facilities. Many of them have died for Gaza.

It is also a tribute to those university students, professors and protesters in the West who paid dearly for their activism and solidarity. It is dedicated to those advocates for justice and peace who did all they could to stop the genocide, and to those who had to endure the repression of institutions that were supposed to protect universal moral principles and rights.

The book is dedicated to the conscientious objectors who were imprisoned for refusing to join the Israeli army, to the brave Israelis who showed their sympathy for the victims of Gaza, to the Jewish activists, to the concerned citizens who protested the relentless onslaught.

A thought also goes to those who dream of a peaceful world, without wars and genocides, to those who dissent, act, protest, and yet have not lost hope, despite the deep sense of despair and impotence that seemed to make all action useless.

Finally, the book is dedicated to the children of Gaza who either lost their lives or had one or more limbs amputated, and to all the traumatized kids who will suffer the consequences of this genocide for the rest of their lives.

CHAPTER 1

The Gaze of Gaza

Peace on earth is not for us.
Does the sky above not see us
or do the crosses on our backs in
the fields of bitter blood
obscure us?

IBRAHIM NASRALLAH (translated by Huda Fakhreddine)

⁛

On October 8, 2023, Palazzo Chigi, the headquarters of the Italian government, which two years earlier had been illuminated in the colors of the Ukrainian flag, was lit up in blue and white in solidarity with Israel. The *Star of David* shone brightly as a symbol of unwavering support. Earlier that day, Israel had launched a large-scale military assault on Gaza, described as the "heaviest conventional bombing campaign" in the history of modern warfare.

After October 7, everything in Gaza became a military target: a building, a hospital, a school, a child, a doctor, a humanitarian worker, a patient. In just six months of conflict, about half of Gaza's homes have been reduced to rubble. The attacks were relentless, indiscriminate, and deadly. As the European Union's High Representative and foreign policy chief, Josep Borrell, pointed out, Gaza was transformed into "one big open-air cemetery" (European Union, 2024).

Despite the horrific scenes unfolding in Gaza, several mainstream Western politicians and opinion leaders remained unimpressed. Their blind support for Israel continued even after the *International Court of Justice* (ICJ), in January 2024, responded to South Africa's legal complaint against Israel with measures to protect Palestinians from "acts of genocide." Actually, there were earlier voices depicting the assault on Gaza as ethnic cleansing or genocidal. Just a few days after October 7, 2023, United Nations (UN) Special Rapporteur Francesca Albanese noted, "In the name of self-defense, Israel is attempting to justify what would amount to ethnic cleansing" (OHCHR, 2023b). Similarly, Raz Segal, an Israeli academic and expert on modern genocide, called Israel's

© ROBERTO DE VOGLI, 2026 | DOI:10.1163/9789004748286_004

attack on Gaza "a textbook case of intent to commit genocide" (Democracy Now!, 2023a).

Allegations that Israel committed the crime of all crimes against Palestinians elicited significant outrage in some privileged and powerful circles in Western societies. In just six months of attacks, Gaza counted 34,535 dead and 77,704 injured, with a further 10,000 missing, presumed buried under the rubble (UNOCHA, 2024a). Yet these figures did not convince U.S. Secretary of Defense Lloyd Austin, who in April 2024 declared: "We have no evidence of genocide" (O'Brien & Gould, 2024). In the same weeks, this sentiment was echoed by two writers who published an article on the website of the corporate libertarian think tank *American Enterprise Institute*. "Israel's actions in Gaza do not meet the legal standard of genocide," they assured (Pletka & Soleimany, 2024).

Similar views were also shared by most mainstream Western politicians. Despite the overwhelming early evidence of Israel's genocidal acts in Gaza, U.S. President Joe Biden said he had "no confidence in the numbers that the Palestinians are using ... I have no notion that the Palestinians are telling the truth about how many people are killed" (Mitrovica, 2023). British Labour Prime Minister Keir Starmer also rejected the idea, claiming to be "well aware of the definition of genocide" and that this explains why he has "never described or referred to [the situation in Gaza] as genocide" (Mulla, 2024). Italian Foreign Minister Antonio Tajani claimed that "Israel has hit civilians" but "genocide is something else" and "one cannot forget what happened on October 7" (ANSA, 2024). Unlike other Western leaders, when Donald Trump became president again, he acknowledged that "a civilization has been wiped out in Gaza", but openly declared his full support for it (Middle East Monitor, 2025a).

It is worth noting that most Western politicians have been less ambiguous in assessing cases of mass atrocities committed by countries not allied to the *North Atlantic Treaty Organization* (NATO). After Slobodan Milosevic's ethnic cleansing left mass graves and untold suffering in Bosnia-Herzegovina, Starmer argued with no hesitation that Serbia had committed genocide. Numerous leaders of Western nations were also dead certain that the killing of 7,000 civilians and the displacement of 1 million people in Myanmar were sufficient evidence of a plausible case of "genocide" (Reuters, 2023). Gaza has endured a far worse war scenario, but the same Western leaders who denounced these recent cases of genocide refused to apply the same label to Gaza.

In an interview with *Der Spiegel*, William Schabas, professor of international law at Middlesex University and a leading expert on genocide, was asked whether the term "genocide" applies to the situation in Gaza. When questioned, "Is it actually important what the crime is called?", Schabas responded: "It is often meaningful for victims' groups. Many think that their suffering is discounted without the genocide label." He also explained that there is a practical reason: "The doors of the ICJ, where conflicts between states are litigated, open to many countries only if they invoke the Genocide Convention, for which the court has explicit jurisdiction" (Prosinger, 2024).

Even after more than 19 months of mass killings and repeated blockades intended to starve the entire population of Gaza, many intellectuals, including government lawyers in the United Kingdom, persisted in denying that genocide was actually occurring (Wintour, 2025).

1　　A Denied Genocide

The denial or disagreement over the genocide charge, particularly during the initial stages of Israel's offensive in Gaza, largely relied on claims about the alleged unreliability of mortality data. "Gaza, UN revises downward the number of victims among women and children," was the title of an article that appeared in the Italian newspaper *La Repubblica* in May 2024. According to the editorial, the United Nations (UN) had realized that it had made a mistake in assessing the mortality rate and had halved the estimated number of victims (Basile, 2024). Other Italian media outlets repeated the same refrain echoing more prominent international newspapers. At the time, the claim that the Gaza death toll was overstated had already appeared in global media such as *Fox News* (Eglash, 2024). An editorial in the *Jerusalem Post* went so far as to dismiss official Gaza death tolls as "Hamas propaganda" while accusing the UN of "parroting Hamas in its reporting of the data" (Cope, 2024).

Even professionals who specialize in data analysis have gone to great lengths to question the mortality figures. An article featuring the analyses of Abraham Wyner, a statistician and professor at the University of Pennsylvania, dismissed the figures provided by "Hamas-controlled Gaza Health Ministry" claiming that data were "exaggerated or faked" (Jerusalem Post, 2024). A paper by the *American Enterprise Institute* asserted that "the Health Ministry's numbers are statistically impossible" (Pletka & Soleimany, 2024). Similarly, Gabriel Epstein, a research associate at the *Washington Institute for Near East Policy*, called the Gaza death toll data "totally unreliable" (Epstein, 2024).

Similar views were repeated like broken records by Israeli government officials. In January 2025, with a death toll that surpassed 46,000 people, Israeli Foreign Minister Gideon Sa'ar was asked if there were "too many dead in Gaza" and replied: "No, absolutely not" (Fatto Quotidiano, 2025).

Fortunately, the tragic health situation in Gaza attracted the attention of public health experts and epidemiologists whose profession is to investigate mortality indicators on a regular basis. Their analyses have completely debunked the claims about the unreliability of the data. This is not to say, of course, that the Palestinian Ministry of Health's estimates were perfect. Data collection in wartime is a problematic activity. It is crucial to acknowledge the difficult conditions under which mortality reporting took place in Gaza. The bombardment of critical infrastructure, including healthcare and public health facilities, hospitals, and universities, undoubtedly affected the accuracy of mortality estimates. In such tumultuous times, the counting of deaths is severely compromised.

However, in spite of the extraordinary circumstances of a war zone, the data provided by the Gaza Ministry of Health did not prove unreliable. A study published in *The Lancet* investigated mortality indicators between October 7 and October 26, 2023, and concluded that the Gaza Ministry of Health data was accurate. As the authors put it: "we saw no obvious reason to doubt the validity of the data" (Jamaluddine et al., 2023). The study consisted of examining mortality reports from the Gaza Ministry of Health and comparing them with data from an alternative source, the *United Nations Relief and Works Agency for Palestine Refugees in the Near East* (UNRWA). The findings showed no evidence of inflated mortality rates. In addition, the authors pointed out that the Palestinian Ministry of Health consistently provided accurate mortality estimates during previous conflicts, with minor discrepancies ranging from 1–5% to 3–8% when compared to independent analyses conducted by the United Nations (Huynh et al., 2024).

In July 2024, an editorial appearing in *The Lancet* showed that, although data collection in Gaza had become increasingly difficult, the estimate of 37,396 war-related deaths produced by the Palestinian Ministry of Health at the time was reliable. Moreover, as the researchers of the article explained, armed conflicts have significant indirect health effects to be added to the deaths directly caused by violence. In recent conflicts, indirect deaths caused by hunger, lack of access to healthcare services, lack of water, and other forms of material deprivation ranged from 3 to 15 times the number of direct deaths caused by violence. Applying a conservative estimate of four indirect deaths for every direct death to the 37,396 deaths already reported, they concluded that the conflict had caused a total number of 186,000 deaths. This represented approximately 7–9% of the total population (Khatib et al., 2024).

Later, in February 2025, further analyses published in *The Lancet* confirmed that the Palestinian Ministry of Health was not overreporting, but actually underreporting the number of fatalities by about 41%. This study produced an estimate of 70,000 deaths caused by violence. However, the authors added that their findings underestimated the full impact of the military intervention in Gaza because they did not take into account non-trauma-related deaths caused by disruptions in health care, food insecurity, and inadequate access to clean water and sanitation (Jamaluddine et al., 2025). According to a letter published in the *British Medical Journal*, infectious diseases have been "allowed to run rampant in Gaza," and the appearance of the first case of polio in a 10-month-old baby, partially paralyzed after contracting the virus, in a place that had not had a polio case in 25 years, underscored the gravity of the situation (Sah, 2024).

Despite the overwhelming evidence of the Gaza death toll, published in some of the world's most respected medical and scientific journals, some data-driven skeptics remained unconvinced. Instead of focusing on the overall significance of the Gaza mortality data, even if they were subject to an acceptable margin of error, they fixated on those very margins of error to cast doubt on the reliability of the data. This tendency to prioritize methodological rigor over substantive insight is characteristic of a subset of academics who are more concerned with projecting analytical precision than with addressing the real-world importance of their findings. By privileging minor statistical imperfections over the weight of the data's message, their contributions may be precise, but useless. Or, to put it more bluntly, precisely useless.

2 A Genocide Foretold

Some of those who rejected the ICJ's implied conclusion that Israel was already committing "acts of genocide" during the initial phase of its offensive did not dispute the scale of the health devastation in Gaza but objected to the accusation of "genocidal intent." Legally, genocide requires not only the large-scale killing of a national, ethnic, racial, or religious group, but also a demonstrable intent to destroy that group. Nevertheless, facts and evidence clearly prove that the destruction in Gaza was deliberate, systematic, and meticulously planned.

By early 2024, the organization *Law for Palestine* had documented hundreds of statements by prominent Israeli political, military, media, and public figures that could be interpreted as inciting or endorsing genocidal actions. The compilation included over 500 such statements (Law for Palestine, 2024).

Among them was Prime Minister Benjamin Netanyahu's characterization of Gaza as "the city of evil" accompanied by a warning: "Get out of there now. We will act everywhere and with full force" (Boffey et al., 2023). But for most Gazans, leaving is not an option. They have no way out because Israel has imposed restrictions on their movement for decades.

Other examples were even more clearly genocidal in nature. Revital Gottlieb, a Likud member of the Knesset, openly called for collective punishment: "Bring down buildings!! Bomb without distinction! Flatten Gaza. Without mercy!" Similarly, Moshe Feiglin, head of the Zehut Party, said: "It is not Hamas that should be eliminated. Gaza should be destroyed, and Israel's rule should be restored. This is our land." Amit Halevi, another Knesset member, went further, declaring that there should be "no more Muslim land in the Land of Israel … Gaza should be left as a monument, like Sodom." And in one of the most extreme comments, Amichai Eliyahu, Israel's Minister of Heritage, suggested "to drop an atomic bomb on Gaza" (Law for Palestine, 2024).

These incitements unmistakably convey a genocidal intent, making it clear that the main enemy of Israel has never been Hamas, but the entire Palestinian population. As some Israeli soldiers sang while celebrating their indiscriminate attacks on defenseless civilians, "There are no innocent citizens" (Bigg & Gupta, 2024).

It is important to note, however, that calls for ethnic cleansing by high-profile figures in Israel long predate October 7, 2023. As Benny Morris, professor emeritus at Ben-Gurion University, put it in an interview published in January 2019: "If [David Ben-Gurion] was already engaged in expulsion, maybe he should have done a complete job. I know that this stuns the Arabs and the liberals and the politically correct types. But my feeling is that this place would be quieter and know less suffering if the matter had been resolved once and for all." He continued: "If Ben-Gurion had carried … a full expulsion—rather than a partial one—he would have stabilized the State of Israel for generations" (Ofir, 2019).

The question of intent also became paramount in the analysis of epidemiological evidence indicating whether Israel discriminated between Hamas members, civilians, women, and children in its attacks in Gaza. On several occasions, we heard Israeli government officials and army spokesmen claim that they do not target civilians. They also asserted that the level of civilian harm in Gaza is broadly consistent with, and even favorable to, other comparable conflicts in recent decades. Estimates published in *The Lancet* completely debunked these claims. Using multiple mortality data sources to estimate deaths from traumatic injuries between October 7, 2023, and June 30, 2024, a group of scholars found that 59.1% of deaths were among women, children, and older adults. The age and gender distribution of mortality in violent conflict, when considered within a broader framework of evidence, provides insight into the motivations of combatants. A lack of discrimination in killings across age and gender would be numerically reflected in a relatively uniform pattern of risk similar to what the *UN Inter-Agency Group for Child Mortality*

Estimation observed, for example, during the 1994 Rwandan genocide. The study found that the deaths of women and girls in Gaza largely followed this pattern (Jamaluddine et al., 2025).

A report by a London-based organization called *Airwars* corroborated these findings. The study focused on the pattern and intensity of bombardments during the first weeks of Israel's campaign in Gaza, comparing the level of civilian harm with military campaigns documented over a decade of work in some of the world's most intense and complex conflict zones. According to the report, by almost every indicator selected, the harm to civilians in the first month of Israel's campaign in Gaza is unmatched by any other military campaign in the 21st century. The findings showed that, between October 7 and October 31, 2023, the number of civilians killed in Gaza was nearly four times higher than any single month in *Airwars'* records since 2014. In just twenty-five days, the number of children killed was seven times higher than any previous peak recorded by *Airwars* in other war zones. Families were killed together and in their homes in unprecedented numbers: over 90% of women and children were massacred in residential buildings (Airwars, 2023). These figures confirmed earlier UN analyses indicating that 70% of those killed in Gaza were women and children, with 80% of children dying in their own homes (Malik, 2025).

3   "Nowhere Is Safe"

The unprecedented death toll in Gaza reflects the intensity of the indiscriminate bombardments carried out by the Israel Defense Forces (IDF), revealing a deliberate disregard for the principles of distinction and proportionality under international law. Over the course of several months, Israeli attacks have destroyed an extraordinary number of homes and infrastructure facilities, including water and sanitation systems, electrical grids, schools, and health facilities. There has also been widespread damage to bakeries, businesses, places of worship, cemeteries, cultural and archaeological sites, and municipal and judicial buildings.

A report entitled *A Cartography of Genocide* by the research organization *Forensic Architecture* focused on military interventions between October 7, 2023, and September 16, 2024. The analysis documented indiscriminate attacks on a range of civilian and cultural targets, including universities, schools, mosques and key agricultural infrastructure such as fields, orchards, greenhouses, and water wells, as well as facilities for the transport, storage, and distribution of humanitarian aid (Forensic Architecture, 2024).

A comprehensive satellite imagery assessment conducted by the *United Nations Satellite Centre* (UNOSAT) in April 2025 revealed that a total of 174,486 structures in Gaza were affected: 70,436 destroyed, 18,588 severely damaged, and 51,962 moderately damaged. In October 2025, approximately 83% of all buildings in Gaza City were damaged (UNOSAT, 2025).

The scale of the devastation was underscored by Achim Steiner, Administrator of the *United Nations Development Programme* (UNDP), who described the destruction as "unprecedented." More than 50 million tons of rubble covered the territory of Gaza at the time, much of it contaminated with unexploded bombs. Experts estimate that clearing the rubble alone could take up to 20 years (UNOCHA, 2025b).

Several mass graves have been discovered in Gaza. Some of them were located near healthcare facilities. Many people who tragically lost their lives had sought refuge in hospitals, believing they would be safe (McKernan, 2023). This has never been the case. The *World Health Organization* (WHO) strongly condemned Israel's attacks on hospitals and personnel and criticized the Israeli government's orders to evacuate areas of the Gaza Strip, calling them "death sentences" for the sick and injured. The WHO also highlighted the dire circumstances of those in intensive care or on life support, including newborns in incubators and those requiring hemodialysis (WHO, 2023).

According to the United Nations, Israel's assault has also "obliterated Gaza's educational system." This has been characterized as a "scholasticide." Between October 7, 2023 and February 25, 2025, approximately 71% (403 out of 564) of school buildings in Gaza were hit or damaged. At least 612 members of school staff were reported killed and 2,769 were injured. During the same period, over 57 university buildings were destroyed, and more than 190 members of university academic staff were reported killed (OHCHR, 2025d). This amounts to a "universitycide" as well.

Particularly cruel have been the savage attacks on aid delivery, with indiscriminate killings of unarmed children and civilians waiting for food from humanitarian aid entering Gaza. This happened repeatedly. For example, on February 29, 2024, *Doctors Without Borders* stated: "We are horrified by the latest news from Gaza City, where over 100 people were killed and about 750 wounded today ... after Israeli forces reportedly opened fire as Palestinians were waiting to receive food from aid trucks" (Doctors Without Borders, 2024a).

Israel's actions have also forced the displacement of the entire population of Gaza, most of whom are already refugees. These displacements have occurred several times, requiring the rapid evacuation of vulnerable people, including children, the elderly, the wounded and the disabled. The refugees were repeatedly moved from "safe zones" that were never safe and were constantly bombed. In addition, the army killed people who were occupying places and roads that Israel had designated as "safe corridors."

For example, on October 13, 2023, the Israeli army ordered the evacuation of 1.1 million people from northern Gaza. This forced transfer of a civilian population as part of an organized offensive is considered a crime against humanity by the *International Criminal Court* (ICC). Cruelly, after forcing Palestinian citizens to leave northern Gaza within a 24-hour period, the Israeli army then targeted convoys of civilians attempting to comply with the orders, resulting in the deaths of 70 people (Wall Street Journal, 2023a).

With nowhere to flee, no way out of the Strip, the choice for Gazans has tragically boiled down to either enduring a slow death from starvation and disease or facing a quicker demise from relentless bombardment. Giora Eiland, a senior research associate at the *Institute for National Security Studies* and former head of the *Israeli National Security Council,* sounded genocidally prophetic when, on October 12, 2023, he wrote: "Israel needs to create a humanitarian crisis in Gaza, compelling tens of thousands or even hundreds of thousands to seek refuge in Egypt or the Gulf." He also added: "Gaza will become a place where no human beings can exist" (Eiland, 2023).

Despite persistent global protests and repeated appeals by human rights organizations for a ceasefire and unimpeded humanitarian access to Gaza, for months much of the international community remained steadfast in its support for Israel's military actions. As some activists observed, "the only human right left for Palestinians has become the right to die."

4   The War on Children

One of the most tragic images of the Gaza genocide is that of Samia al-Atrash holding the small body bag of her two-year-old niece, Masa Khader. On October 21, 2023, an Israeli airstrike flattened her home in Rafah, killing Masa, her parents Loay Khader and Samar al-Atrash, her sister Lina, and their neighbors.

It took rescuers hours to reach the rubble, where they found Masa's tiny body. The next day, Samia cradled her niece's lifeless corpse. A photographer captured the moment as Samia clung to the child she could not bear to let go. The photo went viral, turning Masa's story into a symbol of Gaza's stolen childhoods.

4.1   *"We Are All Complicit"*

Never in recent history has an army killed as many children in a war zone as Israel has in Gaza.

In an article entitled *When is enough enough?* published in the *British Medical Journal,* three public health experts compared child mortality in recent urban conflicts. Their analysis concluded that civilian deaths and child fatalities in Gaza since last October are unprecedented (Bhutta et al., 2024).

Countless condemnations were also issued from UN agencies and international organizations. James Elder, a spokesman for the *United Nations Children's Fund* (UNICEF), said that Gaza had become "the real-world embodiment of hell on earth for its one million children" (UNICEF, 2024). Along the same lines, UN Secretary General Antonio Guterres observed that Gaza had become "a graveyard for children" (BBC, 2023b).

The Israeli military has repeatedly claimed that it does not target children, insisting that civilian casualties are an inevitable consequence of Hamas's embedding within the population. However, a survey of 65 medical volunteers, reported by *The New York Times,* found that 44 of them had personally witnessed multiple cases of pre-teen children shot in the head or chest (Sidhwa, 2024).

The testimony of Professor Nizam Mamode, a retired surgeon who volunteered in Gaza hospitals and later addressed the British Parliament, described the horror of Israeli drone strikes deliberately targeting injured children. "Sixty to seventy percent of those we treated were women and children. The majority of the child casualties were very young," Mamode reported. When asked if the targeting was intentional, he answered without hesitation: "Absolutely. There is no question in my mind. The experience of countless health workers confirms this time and time again. I have worked in conflict zones all over the world, and I have never seen anything on this scale. Ever" (BBC, 2024).

The atrocities against Gaza's children did not end with bullets and bombs. As a *British Medical Journal* editorial noted, Israel also used starvation as a weapon of war (Sah & Dawas, 2024).

On October 8, 2023, Israel imposed the first total embargo on Gaza, cutting off food, medicine, electricity, and fuel. Within months, Gaza was plunged into a humanitarian catastrophe, with no access to basic needs, including clean water. By March 2024, according to WHO and UNICEF, more than 50,000 children in Gaza were already suffering from severe malnutrition, and dozens had died of starvation. Over 1 million Gazans faced "catastrophic" food insecurity, with children most at risk of malnutrition-related diseases (UNOCHA, 2024b).

About a year later, after a brief ceasefire allowed some aid and relief to enter Gaza, Israel imposed a new total blockade. According to UNICEF, more than

9,000 children have been hospitalized for treatment of acute malnutrition since the beginning of 2025. Diarrhea, particularly deadly among malnourished children under five without access to safe water and sanitation, accounted for one in every four cases of illness recorded in Gaza (UNICEF, 2025). As Mike Ryan, the executive director of the *WHO Health Emergencies Programme*, noted, "We are starving the children of Gaza ... We are complicit. We are causing this, you, us and everyone who does nothing about it" (The Journal, 2025).

However, some disagreed. Jonathan Turner, leader of the British advocacy group *UK Lawyers for Israel*, is one of them. In May 2025, amid growing warnings of famine in Gaza, he criticized a paper published in *The Lancet* that showed that the official death toll in Gaza was significantly underestimated because it excluded deaths from nonviolent causes such as malnutrition. Turner objected to the study's conclusions, claiming that it "ignored factors that may increase average life expectancy in Gaza bearing in mind that one of the biggest health issues in Gaza prior to the current war was obesity" (Siddique, 2025).

4.2   *The Highest Number of Child Amputees Per Capita in the World*

In addition to causing the highest child death toll in modern warfare, Israel set another grim record of pediatric horror by making Gaza the place with the highest number of child amputees per capita on earth (UNICEF USA, 2025). According to the Palestinian Ministry of Health, in October 2024, children accounted for 800 of the 4,500 amputations documented during that period (UNOCHA, 2024c).

An analysis by *Save the Children* declared that "Gaza is redefining war injuries," though it presumably meant Israel. The report showed that "in the first 11 months of 2024, at least 5,230 children sustained injuries requiring significant rehabilitation support, which is inaccessible due to continuous attacks by Israeli forces on hospitals and medical workers and restrictions on the entry of critical supplies" (Save the Children, 2025). As a *New Yorker* editorial asked, "More than a thousand children injured in the war are now amputees. What does their future hold?" (Griswold, 2024).

The genocide not only affected children's physical health, but also their mental health. More than one million of them have experienced devastating trauma. A study of children living through the conflict in Gaza found that 96% believed their death was imminent, 79% had nightmares, 73% showed signs of aggression and 49% wanted to die (War Child, 2024). Underlining the gravity of the situation, Helen Pattinson, the executive director of *War Child UK*, said: "Gaza is one of the most terrifying places in the world for a child to live" (Borger, 2024).

In addition to enduring amputations, starvation, unbearable living conditions, relentless bombardment, and the lasting trauma of genocide, the children of Gaza face yet another profound tragedy: the loss of their parents. The Palestinian Ministry of Health reported that, as of early April 2025, 38,495 children in Gaza have lost one or both parents (Al Jazeera, 2025b).

Few images capture this heartbreak more vividly than that of eight-year-old Zain Mhanna, found sleeping beside his mother's grave. For a whole week, night after night, he curled up on the same patch of earth in search of an affection he would never feel again.

When asked why he slept there, Zain replied, "Because I want to sleep in my mom's arms."

5   History as if It Began on October 7

To justify Israel's massacres, many Western politicians and media spokespeople repeated the argument that Israel's military actions were merely a response to Hamas's October 7 attacks. According to this perspective, no crimes would have been committed in Gaza if Hamas had not initiated the hostilities.

This could not be further from the truth. Those who now denounce the ethnic cleansing of Palestine two years too late are not alone: we are, in truth, almost all at least 77 years late.

Some institutional voices have had the courage to admit it. As UN Secretary General Antonio Guterres stated at the end of October 2023, "Nothing can justify the deliberate killing, injuring, and kidnapping of civilians—or the launching of rockets against civilian targets. All hostages must be treated humanely and released immediately and without conditions." But then he added: "It is important to also recognize the attacks by Hamas did not happen in a vacuum. The Palestinian people have been subjected to 56 years of suffocating occupation. They have seen their land steadily devoured by settlements and plagued by violence; their economy stifled; their people displaced, and their homes demolished" (UN, 2023).

Israel's military operations in Palestine must be examined in the light of decades of invasion, mass displacement, indiscriminate killings, land annexation, unrestricted bombings, racial oppression, illegal settlement attacks, and the systematic destruction of a people. As outlined in a document submitted to the ICJ by the South African government, recent events must be placed in the "broader context of Israel's conduct towards the Palestinians during its 75 years of 'apartheid', its 56 years of belligerent 'occupation' of Palestinian territory, and its 16 years of 'blockade' of Gaza" (ICJ, 2024). Of course, this provided the context, not the justification for the October 7 attacks on civilians, although the Palestinians have the right to self-determination and to defend their own territory, in accordance with international law.

In 2022, *Amnesty International* published a report entitled *Israel's apartheid against Palestinians* indicating that since 1948 and the mass expulsion of over 700,000 Palestinians from their homes, villages and towns (the "Nakba" or catastrophe), Israel has imposed a system of oppression, domination, segregation and deprivation of social and economic rights (Amnesty International, 2022). Israel has since occupied and exercised total control over Palestine, turning Gaza into what *Human Rights Watch* called an "open-air prison." This includes restrictions on movement and a blockade of basic necessities, with total control over the entry of food, water, electricity, medicine, as well as access to fishing and relations with the outside world (Human Rights Watch, 2022). As underscored by another *Human Rights Watch* report entitled *Israel: 50 Years of Occupation Abuses*, Israel has been guilty of war crimes for over half a century including home demolitions, destruction of villages, abuses and repeated violations of international law (Human Rights Watch, 2017).

In addition, before October 7, 2023, Israel carried out indiscriminate massacres of Palestinian citizens. The Palestinian death toll from the Israeli army's punitive incursions between 2008 and September 2023, such as *Operation Cast Lead, Operation Pillar of Defense*, and *Operation Protective Edge*, exceeds 6,000 deaths (UNOCHA, 2025a). A significant proportion of these fatalities are not just collateral casualties in the process of attacking military targets, but deliberate assaults on defenseless civilians. In a widely read and cited *Jerusalem Post*, Israel's military approach to Gaza was characterized as a "mowing the lawn" strategy (Inbar & Shamir, 2014).

The argument that Israel's military actions are meant to defeat Hamas, prevent rocket fire, and "restore peace" has also been completely exposed by what is happening in the West Bank, an area not under Hamas control. Since October 7, 2023, more than one thousand Palestinians have been killed there (OHCHR, 2025). Hamas does not govern the West Bank, yet the Israeli army and illegal settlers continue to kill, occupy, and destroy. Remove Hamas from the equation, and you still have ethnic cleansing.

In the months leading up to the October 7 attacks, Israel set a new record for developing new settlements in the West Bank. These actions are illegal under international law. More than half a million settlers have moved to the area, many of whom regularly carry out violent acts or terrorist attacks. During the first half of 2023 alone, there were around 600 attacks (Goldstein, 2023). Now, they are countless.

As revealed by internationally renowned journalist Louis Theroux in his acclaimed documentary *The Settlers*, the ultra-nationalist Israeli settlers are very open about what they are up to (Theroux, 2025). They speak with frankness and clarity about their racism, hatred, and outright intention to ethnically cleanse the region of all Palestinian life. The settlers' relentless terrorism against civilian life in the West Bank and their expansion of illegal settlements are carried out with total normality and impunity. Western governments either tacitly support it, remain indifferent, or issue some mild, empty condemnations to pretend they care about international law. As Palestinian peace activist Issa Amro, who was featured in the documentary, explained, after sharing footage of armed soldiers and settlers raiding his home in Hebron, such attacks persist because Israel feels shielded by the unconditional backing of Western governments, especially the United States.

6   The West's Heart of Darkness

When the ICC issued an arrest warrant for Russian President Vladimir Putin in March 2023, it was widely praised in mainstream Western intellectual and political circles (BBC, 2023a). But when the same court issued a similar warrant for Israeli Prime Minister Benjamin Netanyahu, accused of far worse and more extensive war crimes, the reaction of the international community was very different. Silence, legal reservations and even cases of indignation replaced widespread applause and political support.

In just six months of attacks on Gaza, the Israeli military killed over twenty times as many Palestinian children as the Russian army killed in Ukraine in more than two years of invasion. Despite this staggering difference, White House Press Secretary Karine Jean-Pierre stated that the United States does not support the ICC's investigation of Netanyahu, claiming, "We don't think they have the jurisdiction" (Dadouch & Westfall, 2024). Even more disturbing was a letter from Republican senators directly threatening the ICC: "Target Israel and we will target you. You have been warned" (M. Berg, 2024). With Trump's electoral victory, the situation worsened further.

Why such blind support for Israel even in the face of its genocidal activities in Palestine?

Some commentators suggest that Western complicity in the Gaza genocide and loyalty to Israel is rooted in historical guilt over the Holocaust. Others emphasize the shared values of two "white civilizations" and argue that Israel is nothing more than the West's outpost in the struggle against Arab-Muslim civilization.

Whatever the historical or geopolitical motives, the West's selective application of justice and compassion has become apparent even to the most colonized minds. The hypocrisy is particularly egregious because the West has long branded itself as the global defender of freedom, democracy, and dignity. The genocide in Gaza, however, has forced Europe and the United States to confront not only their present complicity, but also their colonial, imperialist past.

In Joseph Conrad's novel and masterpiece, *Heart of Darkness*, the writer exposed the moral corruption at the heart of empires that justified conquest in the name of civilization. Through the journey of the protagonist Marlow and his encounter with the enigmatic ivory trader Kurtz, the novella exposes the hypocrisy of imperial rhetoric that cloaks greed and violence in the guise of progress and enlightenment. As Marlow puts it, "The conquest of the earth, which mostly means the taking it away from those who have a different complexion or slightly flatter noses than ourselves, is not a pretty thing when you look into it too much" (Conrad, 1988).

But there is an even more important and devastating message in Conrad's novel, embodied by the descent into madness of Kurtz. In the novel, he becomes a symbol of unchecked power and moral decay. His infamous psychological degradation reveals how imperial and colonial systems not only exploit native populations but also corrupt the colonizers themselves.

Today, Israel's devastation of Gaza mirrors the same atrocities that European powers committed in Africa, India, and the Americas. As with past acts of European colonialism, this genocide is also deemed necessary by the colonizers. Like their predecessors, the modern colonial settlers in Israel justify genocide with a web of convenient, self-serving rationalizations such as defending democracy, protecting peace, and waging a necessary fight against evil.

These delusions, however, are beginning to unravel as the destruction of Gaza forces an inevitable moment of truth. Gaza's gaze is not only a diagnosis of Israel's collective hatred of Palestinians. It also reveals the psychopathological conformity of Western institutions that have normalized a genocide. Here, the pretense of defending human rights and dignity collides with unconditional support for mass atrocities.

The tragic images emerging from Gaza have already left an indelible stain of blood on the West's conscience. The undeniable reality of innocent lives crushed beneath the rubble demands not only a reckoning with our self-professed values, but also with our role in enabling such tragedies. Out of these horrors, the very sense of Western identity lies in ruins. As one slogan declared: "Under the rubble of Gaza, the West buried its own soul."

Selectively Empathic

If a life is not grievable, it is not quite a life; it does not qualify as a
life and it is not worth a note.

JUDITH BUTLER

∴

On Christmas Day 2024, as families across Europe and the United States gathered around dinner tables, exchanging words of love, joy, and celebration, Gaza was under a heavy bombardment. Israeli airstrikes targeted yet another hospital, and in the freezing tents of refugees, one more child was reported dead from exposure.

The lack of international intervention to stop the massacres in Gaza exposed a blatant double standard, especially since Western institutions took a completely different approach during the Russian invasion of Ukraine. While innocent civilians in Gaza were being bombed, starved, amputated and torn to pieces, life in the West went on as usual. While their tax dollars continued to fund a massacre of countless children thousands of miles away, most citizens of Western liberal democracies went about their daily lives, working, shopping, tucking their children into bed and kissing them good night.

How could this happen? Why was nothing done to stop the genocide in Gaza?

James Wilson, producer of *The Zone of Interest*, a film about the daily life of a Nazi soldier's family living next to the Auschwitz concentration camp, observed: "There are obviously things going on in the world, in Gaza, that remind us strongly of the kind of selective empathy where there seem to be groups of innocent people being killed that we care less about than other innocent people" (Roberts, 2024).

1 Empathy and Its Discontents

Inas Abu Maamar was immortalized in a moment of unspeakable grief, cradling the lifeless body of her 5-year-old niece, wrapped in a white shroud, killed alongside her mother and sister by yet another bomb dropped in the name of "Israel's right to defend itself." The picture, which won a *World Press Photo* award, seemed to confront humanity with a silent but evocative question: "Do you feel our pain?"

The word "empathy" refers to the ability to see the world through the eyes of another person. Although, as some scholars argue, there are probably as many definitions of the term as there are researchers working on the subject, the concept comprises at least three dimensions: cognitive, emotional, and motivational (Riess, 2017).

The first, the cognitive side of empathy, or perspective-taking, involves understanding another person's thoughts, attitudes, and emotions. The second, the emotional side of empathy, involves sharing another person's feelings, also known as emotional matching or even emotional contagion. This refers to the idea of feeling sad when others are sad or happy when they are filled with joy. There is also a third aspect of empathy, the motivational component, often called empathic concern or compassion, which stems from the statement "your pain is my pain," which in turn can lead to a desire to help and alleviate people's suffering (Wang et al., 2023).

We know that most people are capable of empathy, and not just from philosophy and psychology.

A team of neuroscientists led by Giacomo Rizzolatti at the University of Parma discovered a remarkable class of brain cells known as mirror neurons, popularly renamed as "empathy neurons" (Rizzolatti & Craighero, 2004). These neurons were first observed in monkeys, where researchers found that certain cells in the premotor cortex activate both when the monkey performs an action and when it watches another monkey perform the same action. Since then, rudimentary mirror neuron systems have been found in other animals including elephants, dolphins, and dogs (de Waal & Preston, 2017; Range et al., 2009). In humans, researchers have discovered that similar neurons can mirror not only physical actions, but also emotional experiences. In effect, observing another person's suffering can trigger neural responses in the same areas of the brain that process our pain (Ferrari & Rizzolatti, 2014). This suggests that the common phrase "your pain is my pain" is not merely metaphorical; it is grounded in measurable neurological processes.

The discovery of mirror neurons has attracted considerable scholarly and public attention, eventually becoming a highly publicized and, some argue, overhyped concept in psychology and popular culture. The discourse on empathy has divided into two main perspectives. One camp sees empathy as an essential part of our moral compass, a prerequisite for prosocial behavior and the motivation to care for others.

The other camp challenges this view, arguing that empathy is neither inherently beneficial nor universally desirable; in some cases, it may even produce harmful consequences. Which of these perspectives appears more convincing?

Among proponents of empathy's moral role, Frans de Waal, author of *The Age of Empathy: Nature's Lessons for a Kinder Society*, argued that empathy is an essential component of our social nature (de Waal, 2009). Similarly, in *The Science of Evil: On Empathy and the Origins of Cruelty*, psychologist Simon Baron-Cohen draws on neuroscience and psychology to examine individual differences in empathic capacity and posits empathy as a crucial factor in understanding both prosocial and antisocial behavior (Baron-Cohen, 2012). Complementary to these analyses, Jeremy Rifkin's *The Empathic Civilization: The Race to Global Consciousness in a World in Crisis* offers a visionary synthesis of history, economics, psychology, and neuroscience, arguing that empathy is humanity's most important resource for addressing the most pressing global challenges of modern society (Rifkin, 2010).

Some, however, disagree. Paul Bloom's *Against Empathy: The Case for Rational Compassion* presents one of the most prominent critiques, arguing that empathic responses often lead to biased, shortsighted, and morally questionable choices (Bloom, 2017). Bloom contends that empathy is easily manipulated and misdirected, and advocates instead for a more rational, principle-based approach to moral judgment. Similarly, Fritz Breithaupt, in *The Dark Sides of Empathy*, challenges the assumption that empathy is inherently virtuous (Breithaupt, 2019). On the contrary, he shows how this psychological quality can exacerbate in-group bias, moral distortion and contributes to emotional exhaustion.

Critiques of empathy have found sharp resonance in contemporary assessments of social and political affairs. Consider how prominent "progressive" politicians have employed the language of empathy in public discourse, often to build consensus and consolidate political support. These lofty appeals, couched in moral idealism, can actually mask manipulative motives. Barack Obama, for example, famously declared, "The biggest deficit we have in our society and in the world right now is an empathy deficit" (Bloom, 2017, p. 16). Yet, this statement stood in stark contrast to his administration's launch of deadly drone campaigns (Scahill, 2016) in Yemen, Somalia and Pakistan, operations that killed some combatants but also thousands of innocent civilians (Tayler, 2013). While his rhetoric evoked moral inspiration, his policies told a different, deadly story.

Conservative politicians have also invoked empathy while enacting policies that betrayed it. George W. Bush, who initiated the invasion of Iraq, resulting in the deaths of hundreds of thousands of civilians following false claims about weapons of mass destruction, campaigned under the banner of "compassionate conservatism." Following the devastation of Hurricane Katrina, he assured the public: "I understand the destruction. I understand how long it's going to take. And we're with you" (Shogan, 2009). Yet, the response that followed prioritized private contracts and profit over the welfare of displaced Americans. Beneath the veneer of compassionate concern, his administration's policies aligned more with the interests of the ultra-wealthy than with the needs of ordinary citizens.

The instrumental use of empathy is not limited to politics. The corporate world, too, has increasingly adopted empathy as a rhetorical device, one designed to enhance profitability and brand loyalty. In her essay *Decolonizing Empathy: Thinking Affect Transnationally*, Carolyn Pedwell (2016) critiqued the commodification of empathy, particularly in business literature. In particular, she examines *Wired to Care: How Companies Prosper When They Create Widespread Empathy* by Dev Patnaik and Peter Mortensen, where the authors argue that corporate success hinges on a company's ability to "walk in someone else's shoes" (Patnaik, 2009).

The essay showed that, like politicians, corporations often exploit empathy as a strategic tool to humanize their image, appeal to citizens as consumers, and foster emotional allegiance. In these cases, empathy is reduced to a marketing tactic, devoid of substantial ethical commitment. Even more troubling is the ideological undercurrent in books like *Wired to Care*, when the authors suggest that corporate empathy can supplant state regulation. This line of thinking promotes a form of "neoliberal empathy", where private enterprise is cast as morally self-sufficient, even when it may erode workers' protection, environmental standards, fair taxation, and equitable wages.

Given such widespread co-optation, by both politicians across the spectrum and corporations seeking profit, it is no wonder so many have grown skeptical of the word "empathy" itself.

1.1   *Crimes of Solidarity*

Skepticism about empathy has been reinforced not only by its association with manipulative speeches but also by its arbitrary invocation in rhetorical battles between opposing factions. A striking example is its misuse in the context of criminalizing solidarity with Gaza.

Since the beginning of the most recent onslaught, many commentators and supporters of Israel's military actions have accused those who have expressed solidarity with the Palestinians of lacking empathy for Israel. Expressions of solidarity with Gaza have even been labeled and stigmatized as support for terrorism. In an article entitled *Why So Many Struggle To Empathize With*

Israel, the author insisted that calls for a cease-fire to stop the massacre of civilians in Gaza showed a "peculiar lack of empathy" for the Israelis killed in the October 7 attacks (Hannah, 2023).

Another piece titled *Empathy Gap*, written by Ksenia Svetlova, a nonresident senior fellow at the Atlantic Council's Middle East Program and a former member of the Knesset, argued that while both Israelis and Palestinians lack empathy for the other, "the current lack of empathy" stems "from the monstrosities of October 7" (Svetlova, 2024). Apparently, what happened in the preceding 77 years did not deserve to be mentioned as a possible cause of this empathy gap.

One of the most surreal cases of criminalization of solidarity came in the case of *Ms. Rachel*, real name Rachel Griffin Accurso, a children's educator and YouTube creator known for her *Songs for Littles* videos, which have received over ten billion views. Unable to remain silent about the genocide of children in Gaza, she decided to share widely circulated images of malnourished Palestinian children and other shocking photos from Gaza. This, however, sparked a response from the U.S. group *StopAntisemitism.org*, which accused her of ignoring "the suffering of Israeli victims, hostages, and Jewish children." The group formally asked the U.S. Department of Justice to investigate whether she was "being remunerated to disseminate Hamas-aligned propaganda" and acting as a foreign agent, simply for posting evidence about Palestinian children killed and starved by Israel (Gedeon, 2025). It was, quite plainly, a preposterous act of intimidation based on false and politically motivated accusations.

Of course, the charge that solidarity with Palestine equals a lack of empathy for Israel is an argument based on egregiously flawed premises. First, expressing outrage at the systematic bombing, killing, amputating, and starving of Palestinians is not a sign of indifference to the Israeli victims of October 7. This misinterpretation reflects the classic "us versus them" mentality that has made empathy itself a contentious, confusing issue. One can empathize with both Israeli and Palestinian people at the same time.

Second, expressing support for the Palestinians is not an implicit or explicit endorsement of Hamas, as some self-serving interpretations based on a superficial reading of events claim. It is possible to stand in solidarity with the innocent victims in Gaza and refuse to support Hamas. Indeed, most critics of Israel's acts of genocide have unequivocally condemned Hamas's attacks on Israeli civilians on October 7.

Third, another logical fallacy used to stigmatize efforts to show solidarity with Gaza is to label all calls for a ceasefire as acts that endangered Israeli hostages, falsely assuming that the military interventions in Gaza were aimed solely at securing their release. In reality, the Israeli government has provided ample evidence that Hamas is not its primary target. As explained in an editorial published in the Israeli magazine +972, the history of Netanyahu's support for Hamas is not even a secret. The editorial says, "from sabotaging Oslo [peace accord] to funneling Qatari cash into Gaza, Bibi has spent his career bolstering Hamas to help perpetuate the conflict" (Reiff, 2024b).

Opposition to Israel's assault on Gaza has been interpreted as indifference to security concerns in Israel, on the grounds that Israeli violence is necessary for the safety of Israeli civilians. In reality, the assaults on Gaza have made them far less safe. Israel is now one of the most dangerous places in the world for its inhabitants precisely because of the Israeli government's occupation, violence, apartheid, and crimes in Palestine. The horrors unleashed by the Israeli army will haunt Palestinian civilians and children for the rest of their lives, with the serious risk that these atrocities also sow the seeds of future revenge.

Aside from refuting these narratives, there are compelling reasons to stand in solidarity with Palestine while also showing empathy for all innocent victims including Israeli civilians. As was often emphasized after the Russian invasion of Ukraine: there is an occupier (plundering and bombing a foreign country) and the occupied (struggling to survive an occupation). Israel has illegally occupied Palestinian territory for more than half a century, committing systematic atrocities, blocking aid, and reducing Gaza to an open-air prison. The anguish endured by more than two million people, starved, bombed, and sniped at, cannot be separated from the larger historical context.

The level of death and destruction in Gaza is out of all proportion to what Israel has suffered.

Moreover, those who uphold universal compassion and the principles of the Universal Declaration of Human Rights extend their sympathy first to the victims of genocide, not to its perpetrators.

When it comes to the most innocent victims of conflict, children who bear no responsibility for the crimes of governments, this moral imperative is even clearer. In decades of occupation, Palestinian child deaths number in the tens of thousands. For every Israeli child killed on October 7, dozens of Palestinian children were killed by Israeli forces.

Another solid argument that justifies a firm stance for Gaza is the bias of the Western media, which minimizes the suffering of Palestinian civilians while devoting ample coverage to the plight of Israeli victims. Counterbalancing this narrative and giving voice to the voiceless is an act of social justice.

There is an even more compelling reason to stand against this genocide. As citizens of Western countries, we materially supported Israel's slaughter of Palestinian children in Gaza, not those massacred by Hamas on October 7. Our votes, our taxes, our silence, all have helped enable these deaths. It is right to speak out and act against all attacks on innocent people, but our foremost moral duty is to stop the violence carried out in our name.

1.2   *The Dark Sides of Empathy*

Another critical dimension that challenges the notion of empathy as a universal moral virtue is the concept of dark empathy. A 2021 study published in *Personality and Individual Differences* introduced the idea of *dark empaths*, individuals who exhibit elevated levels of empathy alongside traits associated with the Dark Triad: psychopathy, narcissism, and Machiavellianism (Heym et al., 2021). The authors argue that this psychological profile may help explain the weak or inconsistent empirical association between lack of empathy and aggression that was observed in previous research (Vachon et al., 2014). Individuals who possess the cognitive capacity to understand other people's thoughts and emotions, and use this insight manipulatively, may not need to be violent or aggressive. In fact, they often present as cold, calculating and emotionally detached individuals.

Research has also explored the relationship between empathy and narcissistic personality disorder, revealing a complex and often paradoxical connection. Individuals with narcissistic traits often present themselves as superior but rely on manipulating others to gain attention and admiration. Interestingly, studies suggest that those with pathological narcissism tend to possess only the cognitive component of empathy, the ability to understand the emotions of others, while lacking compassion (Vachon & Lynam, 2016). This unusual combination of traits serves primarily as a sophisticated means to achieve manipulative, self-serving ends.

Research on dark empaths is consistent with studies examining the more sinister side of emotional intelligence, suggesting that empathy, typically considered a prosocial trait, can also enable emotional manipulation and deception. Individuals with *Dark Empathy* or *Dark Emotional Intelligence* often appear charming, sensitive, and caring on the surface (Davis & Nichols, 2016). Underneath, however, they often act in self-serving ways and possess a remarkable ability to gaslight and emotionally control others.

Research on "successful" psychopaths has also revealed that some individuals possess a high degree of cognitive empathy, or the ability to understand and anticipate the thoughts and feelings of others, without experiencing their emotions (Blair, 2005). This psychological ability enables them to display superficial charm and extraordinary powers of persuasion, even appearing hypnotic or captivating. Such strength lies not in the emotional connection, but in the ability to read people with precision and use that knowledge to manipulate.

The coexistence of empathy with dark psychological traits and even psychopathy, is puzzling. Is empathy not just the opposite of psychopathy? Is empathy not a basic requirement for being "human," a quality that psychopaths lack? Or is empathy just another tool for advancing our self-interest by exploiting other people's emotions?

The key point lies in the value placed on the cognitive side of empathy in the overall definition of the quality. Is cognitive empathy alone, without benevolent emotions for the situation of others, sufficient to make a person an empath? The pro-empathy crowd may disagree. For them, there is no empathy without the emotional side of empathy. True empathy, it is argued, requires both cognitive perspective-taking and emotional attunement. The moral quality and capacity to understand and respond to the suffering of others is key. For this reason, some scholars have preferred to focus on compassion rather than empathy, even though the two dimensions often coexist as part of the same perception and response (Riess, 2017). If this argument is correct, then the dark side of empathy, or "selfish empathy," which may sound like an oxymoron, cannot be considered true empathy, as it lacks the essential emotional component.

One of the most popular arguments leveraged against empathy is the observation that, at the group level, it often becomes divisive rather than unifying. This is particularly evident in times of war or existential threat, when a siege mentality takes hold. In such circumstances, empathy becomes narrowly focused on loved ones and members of the in-group, often leaving little or none for outsiders. In a nation at war, for instance, showing empathy for the enemy may not only be rare, but can also be seen as an act of treason.

Paul Bloom, in his case against empathy, even claimed that we commit atrocities "because of our ability to empathize with others" (Bloom, 2017, p. 1). But is this really the case? Following this logic, the lack of compassion for innocent victims of genocide, such as those in Gaza, would be a consequence of "too much empathy." But something is missing in this train of thought. When this sentiment becomes so tribal, so in-group oriented, that it triggers hate and violence against an entire population belonging to an out-group, can it still be defined as empathy?

It seems that the case against empathy made by critics like Bloom is not truly addressing empathy itself, but rather selective empathy, a constrained, exclusionary version of it. The problem is not an overabundance of empathy then, but a type of empathy that is parochial, prejudiced, and conditioned by rigid group boundaries. It is this selective empathy, not empathy, that makes people indifferent, or even hostile, to those outside their circle of concern. Atrocities and hatred toward others are not driven by too much empathy for the in-group, but by a profound lack of compassion for the out-group. This reflects an inability to transcend personal biases and prejudices, often shaped by national, cultural, or ethnic divisions (Falla et al., 2021).

Bloom is correct to point out that empathy can be a "poor moral guide", but his analysis stops short of interrogating why and under what conditions this is true. When he notes that "whites don't have enough empathy for blacks, and men don't have enough empathy for women," (Bloom, 2017, p.4) he inadvertently highlights the very definition of selective empathy. He identifies the problem yet misattributes its cause.

Instead of dismissing empathy altogether as morally flawed, it is crucial to investigate under what circumstances empathy becomes exclusionary, why some individuals are more prone to selective empathy than others, and most importantly, what factors, social, political, or economic, can foster a broader, more inclusive, universal form of empathy.

2   Compassion for Some

Many in the West could not feel the pain of Inas Abu Maamar as she held the lifeless body of her niece. This is not due to a general inability to recognize, mirror and respond to emotions with care. Rather, it is the difficulty of adopting the perspective of Palestinians, who are perceived as part of an "undeserving" out-group.

The concept of displaying empathy towards one's own group and hatred towards a particular out-group is epitomized by a group of Israelis known as YogiNazis. Rivka Lafair, who lives in the Shiloh settlement in the southern West Bank, is one of them. She describes herself as someone who "thinks outside the box" and works as a facilitator of meetings and group sessions focusing on yoga and personal development. Although she admits to some cognitive dissonance, she has managed to fuse seemingly incompatible values: genocide and love. In her view, there is no incompatibility between being spiritual, teaching yoga and leading personal development retreats while calling for the expulsion and annihilation of one's enemies. As she boldly states: "I love my people with an undying love, and I hate my enemy with an undying hatred ... One does not contradict the other" (Idan, 2025).

But can love and genocide really be reconciled?

At its core, selective empathy is rooted in in-group bias, a deep-seated human tendency to reserve compassion for those we perceive as similar to ourselves. This bias not only fuels favoritism and discrimination, but also dehumanization. It allows people to rationalize harm, justify indifference, and accept suffering when it happens to those outside their social or national boundaries. The result is a deeply fractured moral universe: one in which certain lives are mourned while others are dismissed (Wang et al., 2022).

Surprisingly or not, psychological research has paid little attention to this phenomenon. As of this writing, the phrase "selective empathy" appears in the title of only one short, self-published book that focuses primarily on narcissism (Dong, 2023). Despite its importance and broad applicability, it is scarcely investigated in the peer-reviewed scientific literature. Why this neglect?

One possible explanation is the implicit recognition that all empathy is, to some extent, selective. Empathy is a finite emotional resource and our capacity to feel for others is naturally limited (Cameron et al., 2016) and biased toward those with whom we share close bonds such as family members, friends, and familiar communities.

This limitation, however, must be distinguished from selectivity. A limited capacity for empathy concerns emotional bandwidth, the psychological strain of caring too much or for too many. Selective empathy, on the other hand, reflects moral and social filters: learned biases that determine who is deemed worthy of compassion in the first place. It is not the result of some emotional limitation or exhaustion, but rather an a priori orientation, a predisposition to feel for some and not for others in a fixed, dogmatic manner, and regardless of the actual capacity for empathy.

Another possible explanation for the oversight is psychology's epistemological focus on the individual. Most research on empathy concerns personal traits, mental health diagnoses, or interpersonal behavior. Empathy is mostly treated as a private function, not a social dynamic. Yet selective empathy is not only personal. It is cultural and even political. It can reflect how societies collectively respond to the plight of different groups, and how media propaganda, history, and social norms shape emotions.

A central element of selective empathy is the tacit, mostly unconscious act of deciding who "deserves" our concern. Research shows that we tend to empathize more with those who are familiar, or similar to us (Bruneau et al., 2017). People are also more likely to feel compassion for named individuals with identifiable stories. They are less likely to see the perspective or pain of faceless victims of war, poverty or other types of misfortunes (Small & Loewenstein, 2003).

Studies have also found that when someone is perceived as morally suspect, empathy is often withheld or reversed (Cameron et al., 2022). People sometimes feel satisfaction or indifference when they believe a person's suffering is "deserved" (Berndsen & Tiggemann, 2020). Even the moral pain of "immoral" targets is perceived as less intense or less deserving of empathy compared to that of morally neutral or "good" individuals (Riva et al., 2016).

This phenomenon aligns with the concept of moral circles, which are implicit boundaries we establish to define our in-group and out-group. Within these boundaries, compassion is granted to those who are seen as "victims" and withheld from those seen as "villains" (Crimston et al., 2018). Social norms further reinforce this selectivity; people tend to conform to socially acceptable patterns of empathy, and often feel no empathy at all for those whom society labels as "negative targets." Expressing empathy for a morally negative figure may even result in social penalties, including less respect and approval for the empathizer (Wang & Todd, 2021).

But how do we determine who is morally upright, or who deserves to be in our moral circle?

Although research consistently suggests that our empathy depends on whether the recipient is considered "good" or "bad", it is also important to recognize the psychological and social mechanisms that determine how we perceive "good" and "bad." A basic sense of fairness and respect for the law tells us when it is fair to punish some people and if they deserve to suffer the consequences of their wrongful actions. Whether you are a blood donor, a volunteer in a war zone, a rapist or a mass murderer makes a huge difference.

Yet our sense of justice is often biased, and we make arbitrary judgments about people's moral profiles. There is evidence that individuals may feel less empathy for targets who are judged to be immoral for reasons unrelated to their actual behavior (Brambilla & Riva, 2017; Stellar et al., 2014). Often, they rely on group identity or distorted information. A person's ethnicity, religion, or nationality may unconsciously influence whether other people are seen as victims or villains. In this way, selective empathy distorts justice. It leads us to grieve selectively, to care selectively, and ultimately to protect some lives while abandoning others.

World-renowned psychologist Albert Bandura found that the dehumanization of victims of misfortune plays an important role in moral disengagement. Individuals show less empathy toward those who are perceived as "lesser humans" or devalued as human beings. Empathy is often activated by the perception that a person deserves compassion because she or he belongs to a common humanity. Conversely, if people are portrayed as "savages", "animals" or "sub-humans", there is no longer any need to put oneself in the shoes of others. Victims are degraded and deemed unworthy of compassion (Bandura, 1999).

Cultural moral double standards can lead to selective empathy, dehumanization and even misperceptions of justice and morality. Psychologists have studied these patterns of influence in the context of what they define as the "just world hypothesis," or the belief that people get what they deserve. This belief can be used to justify injustice by filtering reality through the lens that "good things happen to good people and bad things happen to bad people." And of course, "good people" are members of the in-group. While it is a comforting cognitive shortcut, it is also deeply flawed. It enables us to accept suffering as the natural consequence of moral failure, regardless of whether that judgment is accurate.

This moral asymmetry affects everyone and begins at an early age. Loyalty to kin and clan is nearly universal across cultures (Graham et al., 2009). Studies show that children as young as three are more likely to share with siblings and friends than with strangers (Olson & Spelke, 2008). This seems to suggest that people may have an innate, genetic moral proclivity toward socially biased forms of empathy. But does that mean selective empathy is inevitable?

Hardly.

Selective empathy is only one part of our biological makeup. In *The Neuroscience of Fair Play: Why We (Usually) Follow the Golden Rule*, Donald Pfaff argues that people are hardwired to follow the principle, "Treat others as you would like to be treated" (Pfaff, 2007). In a later work, *The Altruistic Brain: How We Are Naturally Good*, Pfaff further claimed that we are biologically predisposed toward spontaneous acts of kindness. According to his research, altruism is our default mode (Pfaff, 2014).

Moreover, findings in neuroplasticity suggest that we are not necessarily locked into patterns of "us versus them" thinking. Our brain is malleable, capable of being reshaped toward new goals and perspectives. While selective empathy may be guided by automatic, parochial instincts, we can consciously modulate our emotional responses. Through deliberate reappraisal, such as reassessing other people's suffering and their personal circumstances, reframing their moral worth, or challenging in-group/out-group divisions, we can shift the focus of our compassion (Wang et al., 2022).

Studies have shown, for instance, that compassion-based practices like Buddhist meditation can significantly enhance empathic responses and promote pro-social behavior (Luberto et al., 2018). In addition, empathy is not a fixed personality trait; it is context-dependent (Zaki, 2014). That means our environment plays a major role in shaping how and to whom we extend empathy. By transforming the environment, culturally, socially, and economically, we can also reshape the boundaries of our empathy.

Throughout history, remarkable individuals, including heroes, sages, and generous souls, have transcended tribal instincts and embodied universal values such as compassion, generosity, acceptance, and forgiveness.

Political scientist Kristen Monroe, in *The Hand of Compassion: Portraits of Moral Choice during the Holocaust*, studied what she called "heroic compassion." She found that individuals who risked their lives to save others often expressed a profound sense of shared humanity, regardless of group boundaries (Monroe, 2006). Their empathy is not selective but "universal" in scope, (Phillips & Ziller, 1997) prompting some scholars to wonder whether such global compassion can, in fact, be cultivated on a wider scale (Ekman & Ekman, 2017).

A powerful example of moral courage is Dr. Izzeldin Abuelaish, a Palestinian physician and three-time Nobel Peace Prize nominee. In 2009, an Israeli bombing took the lives of three of his daughters. For many, such loss would breed hatred and thirst for revenge. But Dr. Abuelaish chose a different path. A Harvard-trained public health expert who had worked in Israeli hospitals treating both Israeli and Palestinian patients, he refused to succumb to vindictive feelings of retaliation. As he powerfully stated in the documentary *I Shall Not Hate*, hatred is "a contagious disease" we must resist (Abuelaish, 2011). Since then, he has continued to advocate for peace, offering a living example of the power of universal compassion in the face of profound injustice and personal tragedy.

3 35 Minutes and 300 Hours of Silence

On October 7, 2023, during the Hamas-led attack on Israel, four-year-old Ariel and nine-month-old Kfir were kidnapped by the Mujahideen Brigades along with their mother, Shiri Bibas. They died in captivity in Gaza. Their bodies were returned on February 20, 2025.

At a press conference at the United Nations Office in Geneva on November 14, 2023, Ofri Bibas Levy, Shiri's sister and Ariel and Kfir's aunt, pleaded with Benjamin Netanyahu, and with the media, not to exploit their deaths for political gain. She asked only for respect, for space to mourn an unspeakable tragedy. She asked for silence (Times of Israel, 2025).

Yet around this tragedy, there was anything but silence. A public blame game erupted. Hamas claimed that Ariel and Kfir were killed by an Israeli airstrike in November 2023. The Israeli government countered that forensic evidence confirmed they were murdered by Hamas. The truth may never be fully known. But one moral fact remains: Ariel and Kfir would still be alive had they not been abducted. Kidnapping children, let alone endangering their lives or killing them, is not only a violation of international law; it is one of the most horrific crimes against humanity imaginable. In fact, it is unimaginable.

The global outcry and grief over the deaths of Ariel and Kfir was more than justified. At a demonstration and commemoration organized for them by the Jewish community in Milan, the president of the organization declared: "This gathering ... is not against anyone; it is a demonstration of solidarity with these poor children" (La Repubblica, 2025).

The Jewish community had asked the Milan City Council to light City Hall in orange in memory of the two children. Mayor Giuseppe Sala refused. In contrast, the Lombardy region lit up the top floors of its headquarters, *Palazzo Lombardia* and *the Pirellone building*, in a gesture of solidarity with the Bibas family.

"All children are equal," said Italian Senator and concentration camp survivor Liliana Segre, who joined the demonstration. Echoing the Jewish community's call, she emphasized: "This is not political: we are talking about children" (La Repubblica, 2025).

Her words provoked a backlash from activists who have been campaigning for more than a year to stop the massacre of children in Gaza. They recalled Segre's repeated refusal to characterize what was happening in Gaza as a genocide. "To say that Israel is committing genocide is blasphemy," she told an interviewer from the Italian newspaper *Corriere della Sera* (Evangelista, 2024).

Many asked: how could such an intelligent and dignified woman, who is supposed to represent the interests of all people subjected to ethnic cleansing, deny the overwhelming evidence of genocide when it comes to Gaza? How could she remain silent in the face of the massacre of 17,121 Palestinian children and their starvation due to the Israeli blockade?

Confronted with such a glaring example of selective humanitarianism, some activists responded with outrage and a mirrored double standard. They asked: "If Jewish communities appear unmoved by the deaths of tens of thousands of Palestinian children, how can they expect us to mourn the Bibas siblings?"

The pain behind that question is real. But it reflects a false dichotomy. Ariel and Kfir were innocent. Defenseless. They were the victims of a terrible injustice. Their deaths deserve the compassion of all people, regardless.

In fact, there is an even deeper reason why everyone, especially those outraged by the lack of empathy for Palestinian children, must have mourned all Israeli children: not to become like those whose empathy stops at the borders of identity. When some people only mourn those they see as part of their own group, the answer is not to replicate that selective mourning for our own group. The antidote to selective empathy is universal justice. It is a commitment to see every child, every person as equal in dignity and worth.

Daniel Levy captured this idea in his powerful remarks to the United Nations Security Council: "One must not forget the names of the Bibas children, Ariel and Kfir, and the circumstances of their deaths. Nor the name of Laila al-Khatib, a two-year-old Palestinian girl who was killed just days ago in her home in the occupied West Bank while eating dinner with her family. Or five-year-old Hind Rajab, bombed and denied medical care, dead with her family. And those lost in the incubators of Gaza's Al-Shifa Hospital."

He concluded: "A minute of silence for each of the Bibas children would be appropriate, as would a minute of silence for each of the more than 18,000 Palestinian children murdered in Israel's devastation of Gaza, many more buried under the rubble. This silence would extend to over 300 hours" (Middle East Monitor, 2025c).

Why We Fail to Feel

The autumn night is falling on the coast of Gaza
The planes are bombing, destroying, destroying
Here the IDF is crossing the line
To annihilate the swastika bearers
In a year there will be nothing left
And we will return safely to our home
We will annihilate them all.

∴

These words, sung by Israeli children as a chilling anthem to the destruction of Gaza, are real (Khatsenkova, 2023). They were shared by *Kan News*, a news outlet of the Israeli public broadcaster (Nur, 2023). The video, produced by *The Civil Front*, a group known for public campaigns supporting the Israeli military, features a chorus of young voices celebrating the annihilation of Gaza.

Shortly after its release, the video was taken down. Indeed, the cruelty of the text is beyond comment. But it would be wrong to blame the little singers for celebrating the massacre of other children. Kids do not necessarily hate one another unless adults teach them to.

In an interview conducted during the 4th International Session of the Russell Tribunal on Palestine in New York City in November 2012, Israeli scholar Nurit Peled-Elhanan addressed how these attitudes are cultivated. Drawing on extensive research into Israeli textbooks, under both left- and right-wing governments, she described a consistent pattern: Israeli children are taught to see military violence as a source of honor and national pride. From kindergarten to graduation, military forces are idolized as heroes, protectors, even divine figures. Jewish identity is elevated as morally superior and as the champion of universal values, while Palestinians are often portrayed as "cockroaches" or "vermin" (Deutsch, 2020).

These dehumanizing characterizations are consistent with public statements made by prominent Israeli figures, many of which were released well before

October 7. In July 2014, parliamentarian and lawmaker Ayelet Shaked agreed with the idea that all Palestinians are "enemy combatants and their blood will be on all of their heads. This includes the mothers of the martyrs." According to Shaked, these mothers "should follow their sons, nothing would be more just. They should go, as well as the physical homes in which they raised the snakes. Otherwise, more little snakes will be raised there" (Daily Sabah, 2014).

1   State of Collective Psychopathology?

Since the beginning of human history, Homo sapiens has shown an enduring ability to commit atrocities and inflict cruelty, and even in recent times, genocides continue to occur. There are countless historical examples of the crime of all crimes, from the ethnic cleansing of the Native Americans to the Nazi Holocaust, from the Armenian genocide to the mass atrocities in Rwanda and Bosnia.

The Holocaust, the quintessential genocide, remains vivid in everyone's memory. Due to the systematic massacres, concentration camps, use of starvation, torture, and medical experimentation on healthy people, it is considered an exceptional genocide.

The mass killing in Gaza is also a unique genocide. Not only because it is the most recent deliberate, systematic, and organized campaign of ethnic cleansing and aid blockage in recent history, but also because it was carried out live. It is the most videotaped genocide in human history. As the Canadian physician, author, and Holocaust survivor Gabor Maté, who lost his maternal grandparents in a concentration camp, observed: "It's like we are watching Auschwitz on TikTok" (Middle East Monitor, 2024d).

The unfolding of genocide in an era that is supposed to be a pinnacle of civic progress and civilization is shocking. The behavior of people who directly commit acts of indiscriminate, remorseless violence and cruelty is often likened or portrayed as psychopathic. But what does psychopathy mean?

Although there is no official definition of psychopathy in the *Diagnostic and Statistical Manual of Mental Disorders* (DSM), the concept overlaps with the diagnosis of *Antisocial Personality Disorder* (ASPD). This mental condition is characterized by a complete lack of empathy, which removes any moral constraints individuals feel during interactions with others. It is also accompanied by an inability to be honest (Skeem et al., 2011).

Facts are unimportant to psychopaths. They are habitual liars, often even unaware of the extent of their deception. Lying is so ingrained in their behavior that the question of sincerity is almost meaningless. Psychopaths can even pass lie-detector tests. Although they cannot feel emotional empathy, they can be very skilled at putting themselves in other people's shoes through cognitive perspective taking. In fact, they often learn to mimic empathy and use this ability to deceive others and advance their own goals. The empathy that psychopaths feel, if we can call it that, makes them effective manipulators, and although they are immune to guilt, they excel at making others feel guilty.

The most internationally accepted and used measure of antisocial personality disorder in the scientific literature is the *Hare Psychopathy Checklist*, developed by Robert Hare, one of the world's leading authorities on this syndrome. The Checklist has four dimensions. The first refers to "interpersonal factors" such as superficial charm, grandiosity, and manipulativeness. The second relates to "affective factors," including lack of empathy, callousness, and emotional detachment. The third concerns "lifestyle" such as impulsivity, irresponsibility, need for stimulation, and a tendency toward parasitic or exploitative behavior. The fourth and perhaps most recognizable dimension is "antisocial behavior," which includes disregard for social norms, criminal behavior, and a lack of responsibility (Skeem et al., 2011).

Although the operationalization of psychopathy is useful to understand this complex phenomenon, the term has also been used loosely and sometimes meaninglessly. A notable example is the op-ed piece by "influencer" Jordan Peterson in which he labeled pro-solidarity student demonstrators who opposed the Gaza genocide on U.S. campuses as "sanctimonious psychopaths" (Peterson, 2023). The American psychologist clearly and deliberately abused the term psychopathy to stigmatize traits that are exactly the opposite of those found on the *Hare Psychopathy Checklist*. One trait of these students, for example, is empathy toward innocent women and children. Another is guilt for being indirectly responsible for their massacre.

The term psychopathy has mostly been applied to specific individuals. However, there is no reason to assume that such a syndrome, when it affects many people at the same time, cannot also be applied to understand collective behavior. After all, what happens to a nation that commits genocide against another people needs to be understood not only in terms of individual traits, but also on a collective level. To understand psychopathy, context is very important. Although Robert Hare once wrote, "If I hadn't studied psychopaths in prison, I would have done it in the stock market," (Watts & Lilienfeld, 2016) an important setting for studying this condition might also be societies that support and perpetrate mass atrocities.

Historians, sociologists, and anthropologists will likely study Israel to understand how the systematic bombardment and starvation of a people became a plausible policy. At least three types of information can shed light on some key political and societal aspects that enabled the genocide in Gaza: a) the testimony from soldiers who carried out mass atrocities; b) the statements from political and military leaders who gave them the orders; c) the widespread public support for military and government actions.

1.1   *"The Most Moral Army in the World"*

"The IDF should be nominated for the Nobel Peace Prize," wrote Michael Gove, a retired British politician who served in the governments of David Cameron, Theresa May and Rishi Sunak (Gove, 2025). The statement sounds like a provocation, but Gove is not an isolated voice. Many in the democratic West strongly believe that Israeli military forces deserve to be praised for the genocidal operations in Gaza.

The plausibility of this opinion, or perhaps delusion, completely collapses under the weight of overwhelming evidence of the consistent and systematic targeting of civilians by IDF soldiers. Despite the repeated claims by Israeli officials that their military actions are directed solely at "terrorists" and Hamas combatants to ensure "peace," the reality on the ground in Gaza tells a very different story. As stated in the May 2025 press release from the United Nations Human Rights Office of the High Commissioner: "No one is spared—not the children, persons with disabilities, nursing mothers, journalists, health professionals, aid workers, or hostages" (OHCHR, 2025c).

Perhaps the most damning evidence that the IDF committed systematic and deliberate war crimes consistent with the psychopathic personality disorder can be found in the bodies of children arriving at hospitals with sniper-inflicted wounds to the head, chest, abdomen or testicles. Or were they all hit by mistake?

Dr. Irfan Galaria, an American plastic and reconstructive surgeon who volunteered with the humanitarian medical organization *MedGlobal* in Gaza, recounted his experience in a *Los Angeles Times* op-ed: "I stopped keeping track of how many new orphans I operated on," he wrote. "After surgery, they're filed away somewhere in the hospital, and I don't know who's going to take care of them or how they're going to survive." One moment in particular haunted him: a group of children, aged between five and eight, were rushed into the emergency room by their parents. "All had single sniper shots to the head", said Galaria. "These families returned to their homes in Khan Yunis, about 2.5 miles from the hospital, after the Israeli tanks withdrew. But the snipers apparently stayed behind. None of these children survived." For Dr. Galaria, what he witnessed could not be described as war. "It was annihilation," he said (Galaria, 2024).

Other medical professionals have corroborated these accounts. In *The Guardian*, Dr. Vanita Gupta, an intensive care physician from New York who volunteered at the *European Hospital* in Gaza, recounted seeing three critically injured children brought in within minutes of each other. Their families said they had been playing in the street when they came under fire, with no other hostilities in the area. Canadian physician Dr. Fozia Alvi, who also worked in Gaza, documented sniper-inflicted brain wounds in children as young as seven. For Dr. Alvi, these are not isolated incidents or tragic mistakes. Miranda Cleland of *Defense for Children International* Palestine agrees. In her opinion, the shooting of Palestinian children by trained Israeli snipers is not accidental. It is routine practice (McGreal, 2024b).

There are countless more horror stories of the IDF carrying out atrocities against children and random civilians in Gaza. Many of them have been carefully listed in a comprehensive document written by historian and professor Lee Mordechai from the Hebrew University of Jerusalem. Mordechai has compiled a database of thousands of videos, photographs, testimonies, reports, and investigations. Among the most heinous crimes he illustrates are "a woman with a child shot while waving a white flag," "starving girls crushed to death in a bread line," "a handcuffed 62-year-old man run over, apparently by a tank," and "an airstrike targeting people trying to help a wounded boy" (Hasson, 2024).

There are also numerous photos and videos of soldiers celebrating while bombing entire buildings, cheering, looting, mocking, trying on the intimate clothing of dead and displaced women, playing with dead children's toys, taking selfies, bragging, branding bodies and boasting about sexual violence.

It is important to recognize that the hatred that eventually led to these atrocities preceded October 7. In an article published in Haaretz, psychologist Yoel Elizur, also a professor at the Hebrew University of Jerusalem, documented firsthand accounts from Israeli soldiers, recounting acts of brutality committed more than ten years earlier. According to Elizur, many soldiers become consumed by their intoxicating power over life and death. One soldier reflected: "It's like a drug ... you feel like you are the law, you make the rules. As if from the moment you leave the place called Israel and enter Gaza, you are God" (Elizur, 2024).

Another soldier described how violence became normalized under the command of a new officer: "We went out with him on the first patrol at six in the morning. He stops. There's not a soul on the street, just a little 4-year-old boy playing in the sand in his yard. The commander suddenly starts running, grabs the boy, breaks his arm at the elbow and his leg here. He kicked him three times in the stomach and walked away. We all stood there with our mouths open. Looking at him in shock ... I asked the commander: 'What is your story?' He told me: 'These children must be killed from the day they are born. When a commander does that, it becomes legal'" (Elizur, 2024).

Additional testimonies reveal horrific patterns of routine abuse. One soldier admitted to shooting an Arab man four times in the back from ten meters away, later justifying it as self-defense: "We did this every day" (Elizur, 2024). Another account describes an unarmed man, aged twenty-five years, simply walking down the street: "[He] didn't throw a stone, nothing. Bang, a bullet in the stomach. They shot him in the stomach, and he died on the sidewalk, and we drove away indifferent." One reservist spoke about his fellow soldiers: "I saw sadistic people there. People who enjoy inflicting pain on others ... What was most disturbing was to see how easily and quickly ordinary people can detach themselves and not see the reality right in front of their eyes when they are in a difficult and shocking human situation." A reservist doctor shared a similar reflection: "There is total dehumanization. You don't really treat them like they're human beings ... looking back, the hardest thing for me is what I felt, or what I didn't feel when I was there. It bothers me that it didn't bother me. There is a normalization of the process, and at some point it just stops bothering you" (Elizur, 2024).

In an interview with Pulitzer Prize-winning journalist Chris Hedges, Gabor Maté, author of the best-selling book *The Myth of Normal: Trauma, Healing and Illness in a Toxic Culture*, tried to explain how people become "morally insentient" (Hedges, 2024c). According to Maté, this moral numbness is not innate, but is often rooted in unresolved trauma. He argues that the development of moral awareness is a natural part of being human, but this only occurs when a person is nurtured, loved, and cared for. In contrast, when someone grows up in an environment marked by emotional neglect, violence, or abuse, these moral faculties may fail to develop or even regress.

"If you look at the mass murderers," Maté explained, "they were all severely traumatized children. From Himmler to Hitler to Göring, you know? And their hearts were completely shut down" (Hedges, 2024c). Maté pointed to the work of psychotherapist Alice Miller, author of *For Your Own Good: Hidden Cruelty in Child-Rearing and the Roots of Violence*, who came to a similar conclusion (Miller, 1990). In her research into the lives of Nazi and fascist leaders, she found a common pattern: most of the individuals experienced a very strict and rigid upbringing, often filled with abuse, trauma, violence, harsh punishment, and coldness.

During the interview, Maté also referenced Edith Eger, a Holocaust survivor who lost her family in Auschwitz and later wrote the memoir *The Choice* (Eger

& Weigand, 2017). Eger offered a sobering perspective on the topic arguing that while "we are not all descended from Nazis, we all have a Nazi in us." In other words, we can all turn off our empathy and become coldhearted. This dark potential lies dormant within everyone. Whether or not it is awakened mainly depends on the environment one encounters.

Primo Levi, Holocaust survivor, critic of Israeli policies, and author of the bestselling book *If This Is a Man*, echoed the same disturbing truth. He, too, warned that "under certain pressures, ordinary people can become perpetrators of cruelty" (Hedges, 2024c).

Wartime atrocities leave deep psychological scars not only on the victims, but also on the soldiers who perpetrate them. War, let alone genocide, poisons everything: the land, the people, and the very souls of those who wage it. While some Israeli soldiers seem unaffected by guilt or remorse, others have been deeply troubled by what scholars call "moral injury." This condition "occurs when soldiers act against their values and beliefs or participate as bystanders" (Elizur, 2024).

A striking example is the case of Eliran Mizrahi, a 40-year-old Israeli military reservist and father of four. Deployed to Gaza after October 7, Mizrahi returned home visibly changed. According to a CNN report, he suffered from severe post-traumatic stress disorder (PTSD), haunted by what he had seen and experienced. Although he physically left Gaza, the trauma followed him. Before returning for another period of deployment, he took his own life. As one editorial put it: "He got out of Gaza, but Gaza did not get out of him" (Ebrahim, 2024).

According to another editorial published in the *Times of Israel*, the number of suicides among Israeli soldiers has increased significantly since October 7, 2023. In the 15 months that followed, 38 soldiers took their own lives. This compares to 14 suspected suicides in 2022 and 11 in 2021 (Fabian, 2025). While the increase in the number of reservists deployed during this period must be accounted for, the sharp rise is potentially alarming. In a related report, some 1,600 Israeli soldiers have been diagnosed with combat-related PTSD. Many were deemed unfit to return to active duty, even after initial treatment. These soldiers suffered from a range of debilitating symptoms: uncontrollable shaking, confusion, difficulty concentrating, chronic anxiety, depression, sleep disturbances, restlessness, sudden outbursts of anger, and emotional numbness (Cohen/Walla, 2024).

Yoel Elizur suggests that the moral injury experienced by Israeli soldiers may be particularly severe because of the unique way in which the IDF is perceived in Israeli and Western narratives as "the most moral army in the world." Some soldiers must have struggled with their inner voice, their conscience, and the dissonance between their actions and the ideals promoted by Israeli government propaganda. As one of them admitted, "I felt like … like a Nazi … it looked exactly like we were the Nazis and they were the Jews" (Elizur, 2024).

1.2   *The Cabinet*

The unspeakable war crimes committed by Israeli soldiers were not carried out without orders from above. Calls for the complete annihilation of Gaza came directly from top government officials. How should we characterize a state leadership that carries out a genocide after having openly and repeatedly announced its intentions?

In a provocative piece titled *Israel's Government of Psychopaths*, an independent political commentator advanced a bold hypothesis: that members of the Israeli cabinet responsible for launching the most recent genocidal assault on Gaza exhibit antisocial personality traits (Ettingermentum, 2024).

While it is professionally inappropriate to apply clinical psychiatric labels to public figures without direct evaluation and standardized diagnostic assessment, this does not preclude analyses of behavior, rhetoric and decision-making that cost countless innocent lives. Do the observable behaviors and public statements of some Israeli officials align with characteristics commonly associated with antisocial personality traits?

Let us begin with Itamar Ben-Gvir, Israel's Minister of National Security and leader of the far-right Otzma Yehudit ("Jewish Power") party. Ben-Gvir was previously affiliated with Kach, a Jewish supremacist movement designated as a terrorist organization by both the United States and the European Union. Kach advocated for extending Israeli sovereignty over all of historic Palestine, territories west of the Jordan River, and for the removal of Palestinians and other Arabs from these areas, in pursuit of a theocratic Jewish state.

Not only was Ben-Gvir affiliated with a terrorist organization, but he has also been convicted of criminal offenses at least eight times. According to Dvir Kariv, a former Shin Bet official, Ben-Gvir's criminal record was so extensive that, "we had to change the ink in the printer" whenever he appeared before a judge (Margalit, 2023).

As a piece that appeared in *The New Yorker* put it, "Ben-Gvir built a career on provocation" (Margalit, 2023). Over the years, he has openly championed the ethnic cleansing of Palestinians, incited ethnic hatred and racism, and actively promoted the arming of illegal settlers in the occupied West Bank. In his role as Minister of National Security, he oversees both the Israeli police and prison systems, where reports of torture and abuse are widespread and well-documented. Among his openly genocidal statements prior to October 7 was his call for soldiers to "kill" Palestinians surrendering in Gaza, rather than arrest them, an open endorsement of extrajudicial execution (Middle East Monitor, 2024b).

Another figure exemplifying extremist and antisocial tendencies within the Israeli leadership is Finance Minister Bezalel Smotrich. Representing the more overtly religious and fanatical elements of Israel's ultra-nationalist right, Smotrich is a self-declared "fascist homophobe," (Haaretz, 2023) who has referred to LGBTQ+ individuals as "abnormal" (Karni, 2015).

Smotrich is also a vocal proponent of the "shoot-to-kill policy" against Palestinian stone-throwers (Agerholm, 2017). Despite his fanatical positions, he is far from a fringe figure. Since entering the Knesset in 2015, Smotrich has held several senior roles, including Minister of Transportation and, more recently, Minister of Finance, under which he oversees large parts of the occupied West Bank. In this capacity, he has aggressively advanced the expansion of illegal Jewish settlements, further entrenching Israeli control over Palestinian territory.

Like Ben-Gvir, Smotrich has issued numerous openly genocidal statements. In the aftermath of October 7, he suggested that starving two million Palestinians in Gaza "may be just and moral" if it leads to the release of Israeli hostages (D. Karni, 2024). He also explicitly called for collective punishment actions and the "total annihilation" of cities and refugee camps in Gaza (Conley, 2024). In April 2025, he advocated for a continuation of the total blockade of aid entering the Gaza Strip. "Not even a grain of wheat will enter Gaza," he said (Durgham, 2025).

Yoav Gallant, a retired general and former Israeli Defense Minister, is another influential figure pivotal in the genocide. Although Gallant did not begin his political career on the far right, his trajectory steadily aligned with the extreme policies of the Netanyahu government. As Minister of Construction, he actively supported the expansion of Israeli settlements in the occupied West Bank, a key component of efforts to prevent the establishment of a viable Palestinian state. He has also publicly advocated the complete annexation of the West Bank, effectively rejecting any two-state solution.

Gallant's military record is equally unequivocal. In 2008–2009, he commanded Israeli forces during *Operation Cast Lead*, a campaign in the Gaza Strip marked by military tactics that drew widespread international condemnation. The UN established a fact-finding mission in response, resulting in the *Goldstone Report*, which documented numerous alleged war crimes and potential crimes against humanity, including collective punishment against the whole population and the use of white phosphorus munitions in densely populated civilian areas and the Al-Quds Hospital in Gaza City (UN Human Rights Council, 2009). At the time, Gallant expressed no remorse for these actions, stating only that he regretted not being "tougher."

On October 9, 2023, Gallant made headlines by openly endorsing the use of starvation as a weapon of war: "I have ordered a complete siege of Gaza. There will be no electricity, no food, no fuel, everything will be closed ... We are fighting human animals and we are acting accordingly" (Abu Alouf & Slow, 2023).

The International Criminal Court (ICC) issued a warrant for his arrest.

Like Yoav Gallant, Israeli Prime Minister Benjamin Netanyahu is the subject of an arrest warrant issued by the ICC. As the long-time leader of Likud, Israel's mainstream secular right-wing party, Netanyahu has become the embodiment of what many critics describe as a profoundly antisocial political personality: one characterized by a lack of empathy, an absence of remorse, and a callous disregard for human life, including that of the Israeli hostages taken on October 7. As a relative of one of the hostages said: "Netanyahu's strategy is to allow the hostages to get killed. He is a psychopath who wants to see the war continue" (Confino, 2024). *Haaretz* journalist Uri Misgav agrees: "There has been no greater criminal in Israeli history than Benjamin Netanyahu" (Misgav, 2024).

Netanyahu will go down in history textbooks as the main person responsible for one of the worst genocides in recent history, but his lust for the ethnic cleansing of Palestinians long predates October 7. Throughout his decades-long political career, Netanyahu systematically worked to prevent the creation of a Palestinian state through policies of dispossession, settlement expansion, ongoing human rights violations, and defiance of international law. His political survival has endured despite corruption scandals and indictments, largely due to his aggressive rhetoric, fear-based messaging, and strategic alliances with religious and far-right elements.

After October 7, Netanyahu invoked one of the most frightening symbols from the Hebrew Bible: Amalek. In an openly genocidal speech, he declared his determination to "rid the world of this evil [the Palestinians]," referring directly to the biblical passage in which God commands the Israelites to destroy the Amalekites. The verse, from the first *Book of Samuel*, has often been appropriated by extremist elements to justify violence: "This is what the Lord Almighty says: '... Now go, attack the Amalekites and utterly destroy all that belongs to them. Do not spare them; kill men and women, children and infants, cattle and sheep, camels and donkeys'" (Lanard, 2023).

Itamar Ben-Gvir, Bezalel Smotrich, Yoav Gallant and Benjamin Netanyahu are figures who epitomize the most extreme expressions of fanatical political behavior within Israel's leadership. However, they are merely the most visible edge of a much deeper and broader ideological movement. Ariel Kallner, Knesset member for the Likud, for example, once stated: "Right now, one goal:

the Nakba" (Middle East Monitor, 2023a). The term *Nakba*, meaning "catastrophe," refers to the mass displacement and expulsion of more than 700,000 Palestinians during the creation of the state of Israel in 1948. Similarly, Daniel Hagari, an admiral in the Israeli army and a senior spokesman, publicly stated in reference to Israel's bombing campaign that "the emphasis is on damage, and not on accuracy" (Johnson, 2023).

Israeli Heritage Minister Amichai Eliyahu affirmed: "We must find ways that are more painful to Gazans than death," adding that "dropping a nuclear bomb" on Gaza should be considered "an option." He also made clear that "anyone waving a Palestinian flag should not continue living on the face of the earth" (Arnaout, 2024). Another government minister, Tally Gotliv, called for Israel to use nuclear weapons against Gaza. As she put it, "It's time for a doomsday weapon. Don't destroy a neighbourhood. Crush and flatten Gaza" (Middle East Monitor, 2023b). Amiram Levin, a retired general and influential Labour Party figure on the Israeli left, has also called for entire neighbourhoods in Gaza to be "flattened" (Ofir, 2023). In the same vein, for Yair Lapid, another member of the Israeli Knesset, "the majority of the 12,000 dead Palestinians [at the time] were terrorists." He exclaimed "Good riddance" in reference to their death (Law for Palestine, 2024).

May Golan, the Israeli Minister of Social Equality and Women's Advancement, said: "I am personally proud of the ruins of Gaza, and that every baby, even 80 years from now, will tell their grandchildren what the Jews did" (Bhat, 2024). The Deputy Minister of Finance, Michal Waldiger, speaking before the Knesset, invoked her son, a soldier in the IDF, and declared: "I hope my son will kill anyone who stands in front of him and wants to harm the State of Israel. Including children" (Middle East Eye, 2025).

Even the Israeli President Isaac Herzog, who met Pope Leo XIV during his inauguration ceremony on May 18, 2025, released a genocidal statement. On October 12, 2023, he made clear he does not distinguish between civilians and combatants: "It is an entire nation out there that is responsible. This rhetoric about civilians not aware, not involved, it's absolutely not true. They could've risen up, they could have fought against that evil regime" (McGreal, 2023).

In the context of the staggering death toll and destruction in Gaza, one might ask whether this long list of genocidal statements is sufficient to conclude that the Israeli government exhibits psychopathic traits.

Probably not—at least not to those who diagnose antisocial personality disorders in clinical settings. Laurent Guyenot, editor of *Reseau International,* thinks otherwise. In his view, when it comes to Israel's heartless, genocidal policies and politics toward Palestinians, "we are dealing with a psychopath."

He also added: "Only psychiatrists can explain Israel's behavior" (Guyénot, 2023). Tamir Pardo, the former director of the intelligence agency Mossad, also addressed the Israeli government in psychopathological terms. In his opinion, the government is made up of "lunatic extremists" who are "a lot worse than the Ku Klux Klan" (Middle East Eye, 2023a).

1.3 *The Public*

In the wake of the atrocities committed in Gaza, many liberal commentators and media figures have sought to assign blame mostly to Israeli Prime Minister Benjamin Netanyahu, acting as if the genocide occurred in a political and social vacuum. This narrative, however, obscures a deeper and more disturbing reality: the ethnic cleansing of Gaza has been actively supported by a broad majority of Israeli society.

By November 10, 2023, after 10,000 Gazans had been killed and approximately 10 hospitals attacked, a poll conducted by the *Israel Democracy Institute* and *Tel Aviv University's Peace Index* found that 57.5% of Israeli Jews believed the IDF was using "too little firepower" in Gaza. Another 36.6% considered the force "appropriate." Just 1.8% thought the military response was excessive (Gordon, 2023). By May 2024, following the deaths of more than 36,000 people, including nearly 8,000 children, public sentiment remained supportive of further military action. About 73% of Israeli Jews still viewed the military campaign as either "appropriate" or insufficient because it "did not go far enough" (Smerkovich, 2024). In August 2024, Israel's *Institute for National Security Studies* published a poll finding that 65% of Jewish Israelis opposed prosecuting individuals accused of gang raping Palestinian prisoners (Ofir, 2024). A survey conducted in 2025 found that nearly 80% of respondents supported U.S. President Donald Trump's proposal to forcibly relocate all Gazans to neighboring countries (The Jerusalem Post, 2025).

There is also some anecdotal evidence that contributes to our understanding of public attitudes toward the Gaza genocide. For instance, nationalist protesters have repeatedly blocked humanitarian aid trucks, and some citizens have openly called for Palestinians to starve to death or to be killed (The Grayzone, 2024). Countless viral videos on social media have been made by Israeli civilians, celebrating the starvation and deaths of Palestinian civilians and mocking the suffering.

This collective cruelty is not a new phenomenon. During *Operation Protective Edge* in 2014, which killed more than 2,000 people, including 500 children, polls showed that nearly 95% of Israeli Jews supported the military intervention (Beauchamp, 2014). Not only did some support the massacres, they also took pleasure from the suffering of Palestinians: groups of Israelis gathered on hilltops near the Gaza border to cheer, whoop and whistle as bombs rained down on people in a warzone a few miles away. Some brought bottles of beer or soft drinks and snacks as well as old sofas and garden chairs to enjoy the spectacle more comfortably (Sherwood, 2014).

Almost twenty years before October 7, 2023, Michel Warschawski, director of the *Alternative Information Center* in Jerusalem, warned: "The whole society is sick, terribly sick ... Violence is manifested not only in Israeli politics, but also in everyday interactions at home and on the street." He also added: "The mixture of aggressive nationalism and victimization produces a level of violence within Israeli society that is difficult to measure from the outside. Psychologists and social workers keep warning about this escalation of violence, but their warnings seem to have little effect" (Warschawski, 2004). A *Haaretz* piece published in 2015 observed: "Voters need to keep one eye on the news and one eye on the American Psychiatric Association's manual of mental disorders" (Sarid, 2015).

These are provocative suggestions. But can the behavior of an entire nation truly be considered psychopathological?

Hardly. A clinical diagnosis of a society as a whole would require extensive analysis of the mental status of each citizen, or at least a representative sample of them. Nevertheless, the importance of assessing psychological conditions at the aggregate level cannot be underestimated. Apart from rare exceptions, the discipline of psychology does not seem to address psychopathologies at the collective, national level. An exception is a checklist called *Syndrome of Collective Callous-Unemotional Traits* to indicate the psychological state of some groups, communities, and even nations that are dominated by hatred, intolerance and violence (Athar, 2023). The checklist describes collective patterns of behavior and ideology characterized by the following characteristics:

1. Lack of guilt or remorse;
2. Absolutist belief systems;
3. Imposition of the belief system on others;
4. Violation of basic human rights;
5. Suppression of cultural or entertainment practices incompatible with the dominant ideology;
6. Sadistic control;
7. Sharp "us versus them" thinking;
8. Rewards for ideological conformity;
9. Use of violence to enforce the belief system;
10. Involvement in heinous crimes to maintain group identity.

These traits have been used to analyze totalitarian regimes, such as Nazi Germany, and extremist groups, like ISIS or the Taliban. Can they also help illuminate the dynamics within a society that perpetuates a genocide?

Some would greet even the idea of entertaining such a hypothesis with outrage and contempt. One might argue that dimensions such as a rigid, absolutist belief system (dimension 2) and the institutional imposition of ideology (dimension 3), hardly apply to Israel given its electoral system and civil liberties. But even in this case there are important qualifications. While Israel is often considered a democracy, it is more accurately described as an "ethnocracy" (Yiftachel, 2006). In 2018, Israel passed the Nation-State Law, sometimes referred to as the *Jewish State Law*, which declared that "the right to exercise national self-determination in the State of Israel is unique to the Jewish people" (Wootliff, 2018).

Religious and ethnic nationalism run deep in Israel. Surveys show that 86% of Israeli Jews believe in God or a higher power, (Inbari & Bumin, 2024) and 61% agree that Jews are "God's chosen people," who have the right to live in the "Promised Land" (JPPI, 2024). In addition, 87% express a strong belief in Jewish ownership of the land between the Jordan River and the Mediterranean Sea, (Warnke et al., 2024) while 42% support annexing the West Bank without offering equal rights to the Palestinians (PCPSR, 2024). About 47% of Israeli Jews support "the claim that the [Israeli army] in conquering an enemy city, should act in a manner similar to the way the Israelites did when they conquered Jericho under the leadership of Joshua, i.e. to kill all its inhabitants" (Rapaport, 2025). As Erich Fromm stated, "If all nations would suddenly claim territories in which their forefathers had lived two thousand years ago, this world would be a madhouse" (Fromm, 1958).

When examining dimensions 5, 8, and 9 of the checklist (censorship of nonconforming cultural practices, rewards for ideological conformity, and the use of violence to maintain the belief system), it can be argued that the repression within Jewish Israeli society is not even comparable to authoritarian regimes. However, the conditions in Gaza and the West Bank include policies of forced displacement, extrajudicial killings, colonial settlers' terrorism and systematic destruction of property that constitute clear patterns of ethnic cleansing.

Other dimensions such as the violation of basic human rights (4) and sadistic control (6) are unmistakably present in Israel, but again only when considering policies and behaviors toward the Palestinians. The "us versus them" (7) divide is key. Palestinians, whether in Gaza, the West Bank, or within Israel, are subjected to surveillance, restriction, colonial settlers' random violence and endless humiliation. Among Jewish Israeli citizens the situation is completely different, but even activists and intellectuals who advocate for Palestinian rights often face repression, censorship, imprisonment and violence.

While some of these patterns may or may not fit the criteria of the *Syndrome of Collective Callous-Unemotional Traits*, any generalization should be avoided. A significant minority of Israelis have resisted the dominant policies of occupation and genocide, some at great personal cost. Their courage and moral clarity are the antithesis of the characteristics of the syndrome. Moreover, the hostility, cruelty, and lack of empathy are selectively applied. Within the in-group, Israeli society can be deeply communal, and cohesive. Thus, while elements of the *Syndrome of Collective Callous-Unemotional Traits* may offer a provocative lens through which to view the normalization of state violence and dehumanization of Palestinians, any attempt at diagnosing a collective psychopathology for an entire society is problematic and carries the risk of dangerous generalizations.

2   Selectively Indifferent

The division between "us and them," which channels empathy exclusively to the in-group while dehumanizing the out-group, is not unique to Israel. It is an almost universal feature of human societies, especially those exposed to violence or under threat. In contexts of war or perceived existential danger, a siege mentality often dominates, overriding normative social values and emotional responses. Most European countries and the United States, having been largely spared war on their own soil since the end of World War II, may find it difficult to fully understand the psychological transformations that war imposes on social life.

But history provides ample evidence. During World War I, Italians were taught to despise Austro-Hungarians, just as Austro-Hungarians learned to hate Italians with all their hearts. In World War II, British and American propaganda relentlessly portrayed Germans and Japanese people as subhuman, morally bankrupt, or inherently evil. These are not historical aberrations, but recurring features of warring nations. War inculcates in people a hatred of strangers, to turn foreign others into mortal, immoral enemies. War intoxicates people's minds with attitudes that would normally be considered psychopathological in peacetime.

But how does this "us versus them" mentality develop?

Scientific research offers some answers, but also cautions against oversimplification. Selective empathy results from a cascade of multiple, interconnected determinants. Research shows, for example, that some of the most influential factors that generate selective empathy include not only group membership, but also perceived and actual similarity, emotional connection, judgments of deservingness, and dehumanization.

Selective empathy is also shaped by how we are raised, by our individual moral development, social experiences, and cultural context. A person's sense of justice or injustice, and beliefs about hierarchy and dominance all contribute. These traits are not fixed; they vary widely among individuals, societies, and historical circumstances. In addition, macro-level forces such as religion, ethnocentrism, education, media narratives, and cultural norms, play a powerful role in shaping the boundaries of our empathy.

2.1   *Our Brain on Nationalism*

Of all the factors capable of modulating our empathy toward those outside our own group, nationalism is perhaps the most powerful. It can turn complete strangers into either enemies or allies, depending on psychological and social constructions it embeds deep within society. Nationalism does not just make us forget that we belong to a single human family. It can also suppress our ability to feel compassion for others.

In an article entitled *This Is Your Brain on Nationalism* published in *Foreign Affairs*, biologist and neuroscientist Robert Sapolsky explores the biological underpinnings of group identity (Sapolsky, 2019). Sapolsky explains how the amygdala, the brain's center for processing fear and aggression, activates almost automatically when people encounter faces of different races, reflecting our evolutionary tendency to make quick in-group/out-group distinctions. National symbols, narratives, and shared goals can significantly shape who is included and excluded from our circle of empathy. But Sapolsky also emphasizes that while these instincts exist, they are not predestined. Social and cultural constructs can reinforce, weaken, override or redirect these impulses.

Although psychologists have typically viewed a sense of national identity as a factor that positively influences wellbeing and self-esteem, and have found empirical evidence to support this idea, (Yang et al., 2025) nationalism can become toxic and even deadly. Fostering a strong emotional attachment to one country can create a deep sense of belonging and loyalty. However, this often comes at a price: empathy becomes conditional. It flows easily to those within the national in-group but diminishes or disappears when directed at outsiders. National symbols and even national anthems reinforce identity and cohesion, but they also have the potential to desensitize individuals to the suffering of those perceived as "other."

The aesthetic and unifying effect of flags is fascinating, but it can also mask dark psychological powers. Throughout history, nationalist movements and governments have weaponized these symbols to incite hatred, justify inequality, and perpetuate violence. By narrowing the scope of empathy and reinforcing myths of ethnic or genetic superiority, nationalism has enabled atrocities ranging from war crimes to genocide.

While selective empathy exists in any society, its consequences are particularly acute in the context of Israel's ongoing conflict with the Palestinians. In this case, the dehumanization of the out-group has become not only socially acceptable, but structurally embedded, in education, media, national narratives, and even everyday expressions of patriotism.

Rabbi Yaakov Shapiro, author of *The Empty Car: Zionism's Journey from Identity Crisis to Identity Theft*, argues that Zionism fundamentally redefined Jewish identity from a religious belief into a nationalist and even racial construct (Shapiro, 2020). In his view, nationalists have manipulated the emotions of Jewish citizens persuading them they were a nation in need of a state, a flag, and a military, effectively manufacturing a new, artificial identity (Media Mondo, 2024). "Judaism was stolen by the Zionists for immoral reasons" he writes. He also added: "Jews are neither a race nor an ethnicity, but a faith, a religion. The Zionists, however, see the world through their own hallucinations ... [and] They have succeeded in making much of society believe in those same hallucinations" (Media Mondo, 2024).

In Shapiro's view, this ideological shift required a rigid and artificial "us versus them" binary. One of its most pernicious consequences is the systematic dehumanization of Palestinians. The nationalist identity forged by Zionism demands not only political loyalty, but also emotional detachment from the suffering of the Other. Former Israeli military police officer Gil Hillel offers a powerful example of how this transformation unfolds on a personal level. Reflecting on her time in uniform, she described the psychological desensitization that came with enforcing military rule over Palestinians. "Over time, I became more and more violent," she said. "I went through a crazy transformation, from a very calm and relaxed person to a very violent and aggressive person who took out her frustrations on the only people she could: Palestinians and prisoners." She continued, "I hit more; I was more abusive. I didn't really see them at all. They're invisible people, you don't see them" (McGreal, 2022).

2.2   *The Long Shadow of the Holocaust*

As many have watched the horrific and systematic destruction of Gaza, they have wondered how a nation born out of the trauma of genocide could become the perpetrator of genocide. There is no easy answer. But part of the problem lies in the complex interplay between collective trauma, national identity, and the political manipulation of both. As Yaakov Shapiro and other writers argue, Zionist and nationalist ideologies have channeled the emotional legacy of the Holocaust to forge a society united by shared goals, ethnic symbols, and a deep sense of unity (Shapiro, 2020).

To understand how this trauma has contributed to the escalation of violence, culminating in the genocide of Palestine, it is necessary to place recent events, including the aftermath of October 7, within a broader historical context. This includes the emotions that have shaped Israeli society including fear, hatred, and defensiveness. These collective feelings did not emerge overnight. They have been deeply rooted since the nation's founding and must be understood through the lens of the Shoah, a trauma that left an indelible mark on Israel's national psyche (Moussa, 2018).

The Shoah instilled in the emerging Israeli state a profound sense of existential threat which evolved into a political culture defined by hypervigilance and militarization. Framed by the memory of genocide, Israeli identity became intertwined with a persistent sense of victimhood and survivalist urgency. In this context, empathy has become almost exclusively confined to the in-group. The institutionalization of Holocaust trauma within national memory has not only shaped Israeli military policy but also numbed the collective capacity to recognize or feel guilt for the suffering inflicted on Palestinians. This collective psychological framework distorts perception, making it nearly impossible to see Palestinians as victims. Israel, by defining itself as the ultimate symbol of historical victimhood, renders the suffering of others invisible. Through this lens, all acts of aggression, from the ethnic cleansing of 1948 to the genocidal acts following October 7, are interpreted as legitimate defensive measures (Moussa, 2018).

The long shadow of the Holocaust has produced a society gripped by a chronic "siege mentality," a state of collective paranoia where past trauma is inextricably linked to present conflict. This mindset leaves little room for empathy or compassion toward the out-group. The national consciousness is already saturated with fear, anger, and the desire for retribution, all rooted in the symbolic weight of Holocaust memory. This is epitomized by Netanyahu's chilling declaration that Israel "will forever live by the sword" (Shulman, 2015), a chilling summary of the belief that the country's survival depends on perpetual war.

Zionist leadership has strategically exploited this trauma, along with the obsession with security, and skewed perceptions of victimhood, to craft a historical narrative that blurs the line between the roles of victim and perpetrator. In such a framework, empathy for the out-group becomes subversive. It must be denied, even violently repressed, to preserve national cohesion and ideological purity.

Solidarity is criminalized. Any identification with Palestinian suffering poses a threat to the dominant mindset and emotional conformity. It risks revealing a shared humanity, a mutual victimhood that could dismantle the myth of moral exclusivity. Recognizing this parallel suffering would challenge nationalist ideologies at their core, offering a radically different understanding, one grounded not in separation, but in justice, and the universal capacity to feel other people's pain. The sense of shared humanity that transcends borders and ethnic divisions is the ultimate threat to nationalist ideology. For nationalists, it is a mortal enemy, because it renders their creed of violence, hatred, revenge, and obsessive identity politics not only useless but indefensible, even under the most twisted justifications.

3   The Banality of Complicity

The genocide in Gaza could not have been carried out with total impunity without the blind support of much of the Western international community. In the normalization of genocide, two powerful nations, such as the U.S. and the U.K., played a major geopolitical role. Support for Israel's onslaught remained persistent and unwavering across both center-left and center-right political parties. As Ameen Izzadeen observed in the *Daily Mirror*, some Western leaders who expressed support for Israel and its war crimes have been "promoting and celebrating psychopathic behavior and calling it humane" (Daily Mirror, 2023).

Uncritical support for Israel's military operations has already resulted in some legal repercussions as several Western politicians are now under investigation for complicity. Both U.S. Presidents Joe Biden and Donald Trump played a decisive role in enabling Israel to carry out these atrocities. Likewise, European Commission President Ursula von der Leyen has shown unequivocal support for Israel's actions. On May 22, 2024, the *Geneva International Peace Research Institute* (GIPRI) filed a complaint against her with the International Criminal Court (ICC), stating that: "Reasonable grounds exist to believe that the unconditional support of the President of the European Commission to Israel—military, economic, diplomatic and political—has enabled war crimes and the ongoing genocide in Gaza" (Middle East Monitor, 2024c).

As with previous genocides, the normalization of mass atrocities in Gaza has not been made possible only by some political leaders. It has also relied on the silence and complicity of the media and intellectuals. This is unsurprising. History has shown that genocides are rarely carried out by just a few violent fanatics. When asked by Italian journalist Enzo Biagi, how did concentration camps come into being, Primo Levi replied: "By pretending that nothing is happening" (Fatto Quotidiano, 2014).

Philosopher and Holocaust survivor Hannah Arendt, in *The Origins of Totalitarianism*, explored how the machinery of genocide is usually sustained not only by overt perpetrators, but also by "frighteningly ordinary" people who follow orders, remain silent, or simply look the other way. This observation gave rise to her concept of the "banality of evil": the idea that horrific crimes can be committed by ordinary people, not out of malice, but out of compliance and conformity. According to Arendt, the Holocaust was born not only of genocidal intent but also of a failure to think (Arendt, 1973).

Psychologist Stanley Milgram's famous obedience experiments demonstrated the disturbing ease with which people follow the orders of authority figures and violate their own moral principles. His experiments, conducted in the early 1960s, involved one participant (called "the teacher") who was instructed by an authority figure (the experimenter) to administer increasingly severe electric shocks to another participant (the "student") whenever the student gave incorrect answers to a series of questions. Unknown to the "teacher," the "student" was an accomplice of the experimenter and never actually received shocks. Instead, the "student" acted out scripted responses, including verbal protests, pleas to stop, and expressions of pain. In some cases, the "student" even pretended to lose consciousness as the voltage increased.

The results were striking. Despite the obvious distress of the "student," 65% of the participants continued to administer shocks up to the highest voltage level, labeled "xxx," when instructed to do so by the experimenter. This result held even when the "teacher" could clearly hear the "student's" cries of pain and pleas for mercy. Milgram attributed this behavior to what he called the "agentic state." In this psychological condition, people see themselves as instruments carrying out the will of an authority figure, thereby relinquishing personal responsibility for their actions. They dehumanize themselves. This state of mind allows them to suppress their moral concerns and follow the orders of the authority figure, even when those orders conflict with their ethical standards (Milgram, 1963).

Moral suppression can also happen due to conformity to social norms widely accepted in society and the fear of standing against the majority. The results of Milgram's study not only demonstrated the power of authority in shaping human behavior, but also provided empirical support for Hannah Arendt's concept of the "banality of evil," which suggests that ordinary individuals, when placed in certain contexts, can commit harmful or unethical acts without being inherently evil.

Psychologist Solomon Asch's experiments too contributed to our understanding of these phenomena. Asch studied the effects of group pressure on individual judgment. In his most famous study, participants were asked to match the length of a line to one of three comparison lines. Each group had only one real participant; the rest were confederates who intentionally gave the wrong answer on certain trials. Asch found that many participants conformed to the group's incorrect answer, demonstrating the power of social influence and the pressure to conform, even when the correct choice was obvious (Asch, 1951).

In *Bystanders*, a chapter in the book *Psychology of Genocide: Perpetrators, Bystanders and Rescuers*, Steven Baum explored the mindset of those who refrain from protesting or opposing a genocide. According to the author, bystanders are particularly influenced by social norms and make decisions based on what is culturally appropriate, using various types of justifications and rationalizations for their behavior. It is estimated that bystanders typically make up 50–65% of the total population. Their inaction, silence, and deference to power are key to carrying out collective punishment of another people (Baum, 2008). As Howard Zinn has explained in his book *You Can't Be Neutral on a Moving Train*, history clearly and repeatedly shows that the most terrible acts of cruelty, including war, genocide, and slavery have resulted from obedience, not disobedience (Zinn, 2010).

The normalization of evil experienced in Gaza does not remain confined to Gaza. The lack of opposition to the genocide is already having serious consequences worldwide. As former UN official Craig Mokhiber, who resigned early on to protest the inaction of the international community in preventing Israel's atrocities, made clear: when international law ceases to apply, effectively making apartheid and even genocide acceptable, "we are all in trouble." As he argued, "Israeli proxies are also eroding free speech and human rights across the West…Don't kid yourself. You could be next" (Worthington, 2024).

Many have not yet fully grasped the consequences of what is happening. Gaza is often treated as something very distant from us, something that does not concern us.

Those who think this way are greatly mistaken.

It is not just a matter of saving lives and stopping genocide. What is at stake is humanity itself. If genocide is perpetrated with impunity, and the minimum requirements of humanity are trampled on in defiance of international law, then Gaza will no longer be just a distant place.

Gaza will be us. And in part, it already is.

4   **When the West Incites Genocide**

Western politicians and high-profile figures have not merely remained silent or indifferent in the face of the Gaza genocide. In several cases, they have issued statements that can be regarded as incitements to genocide.

Outside Israel, most genocidal declarations have come from the United States. Among these, the case of Republican congressman Randy Fine stands out. In a social media post, he openly called for the extermination of Palestinians, writing: "Kill. Them. All. There is no suffering adequate for these animals. May the streets of Gaza overflow with blood" (Kennedy, 2024).

During a visit to Israel in support of Operation Swords of Iron, former U.S. Ambassador to the United Nations Nikki Haley was photographed posing beside an artillery shell marked with the words, "Finish them!" (Rashid, 2024).

In another episode, Florida State Representative Michelle Salzman, when asked, "We're at 10,000 dead Palestinians. How many is enough?" responded: "All of them" (Salam, 2023).

Faced with questions about Palestinian children's deaths, Representative Andy Ogles responded, "We should kill them all" (Reid, 2024).

Similarly, Senator Lindsey Graham stated that Israel, in relation to the situation in Gaza, must "do whatever is necessary," including an explicit reference to the US atomic bombings of World War II (Liddell, 2024).

Former Representative Michele Bachmann stated that "It is time that Gaza ends. The two million people who live there are clever assassins. They need to be removed from that land" (Media Matters, 2023).

In January 2025, newly elected U.S. President Donald Trump advocated for the forced relocation of Palestinians to neighboring countries, describing it as a feasible and potentially quick-to-implement plan (Graham-Harrison, 2025).

However, the idea of depopulating Gaza had already surfaced in Italy more than a year before.

On October 10, 2023, Italian journalist Claudia Fusani, speaking on the television program *Restart*, justified the use of collective punishment as a means to drive Palestinians out of Gaza. She stated: "In my opinion, it's right that, rather than dropping bombs, they cut off water, electricity, and gas. I'm sorry, because it means bringing people to their knees, but people currently have an escape route through Egypt, so they should use it and empty Gaza" (RAI 3, 2023).

Genocidal statements have also been heard in the United Kingdom. A Declassified UK analysis found that 180 of Britain's 650 Members of Parliament have accepted funding from pro-Israel lobby groups or individuals during their political careers. One particularly provocative case is that of Lord Ian Austin. Social media posts and screenshots have circulated showing that his X (formerly Twitter) profile once displayed an image with the words, "Let Israel finish the job" (McEvoy, 2024).

The West and the Rest

When the missionaries came to Africa, they had the Bible, and we had the land. They said, 'let us pray'. We closed our eyes. When we opened them, they had the land, and we had the Bible.

Museum of Apartheid, Pretoria, South Africa

∴

"Today we also celebrate 75 years of friendship between Israel and Europe," declared European Commission President Ursula von der Leyen on April 26, 2023. She continued: "We have more in common than geography would suggest, our shared culture, our values, and hundreds of thousands of dual European-Israeli citizens … Europe and Israel are bound to be friends and allies. Your freedom is our freedom" (Israel National News, 2023).

Renowned American comedian Bill Maher echoed the same sentiment: "Israelis look like us" (The Hill, 2024). But what, exactly, are the shared values that bind Israel and the West?

According to Joseph Massad, professor of modern Arab politics and intellectual history at Columbia University, these principles are not democracy, freedom of speech and human rights. Rather, they are values of settler-colonialism, racial supremacy, and hegemonic power, or the ideals through which empires have long advanced and protected their geopolitical interests (Massad, 2024).

Israel and the West share more than ideals, however. They also share weapons, military intelligence, and geopolitical strategies. They are not just allies; they are also partners in crimes.

1 Partners in War Crimes

The genocide in Gaza is not the work of just one army. Rather, it is the result of a vast network of nation-states, private intelligence firms, major investment banks, and multinational arms manufacturers. Some of the leading arms suppliers to Israel are defense contractors including *Boeing, General Dynamics, Leonardo, Lockheed Martin, RTX* (formerly *Raytheon*), and *Rolls-Royce*. Despite growing global calls to halt all weapons transfers to Israel to avoid complicity in war crimes, these corporations have continued to develop and supply lethal equipment for use in Palestine (PAX, 2024).

Several major banks have financial ties to companies that supply military systems to Israel. This makes them complicit. These companies are legally and morally responsible under humanitarian law and international criminal law, but they have consistently ignored ethical standards in favor of their business interests and power alliances (PAX, 2024).

Although corporations and financial institutions prioritized profit over opposing the genocide, Western governments are the most responsible for enabling Israel's war machine. Arms exports from both Europe and the United States to Israel have remained steady and unrelenting, both before and after October 7, 2023. According to a *Transnational Institute* (TNI) report, the United States is Israel's primary military supporter, providing over $3 billion in annual aid. European countries also contribute to Israel's military capabilities. Germany is the largest European arms supplier to Israel, having issued export licenses worth €880 million between 2018 and 2022 (Akkerman, 2024). A BBC analysis based on data from the *Stockholm International Peace Research Institute* (SIPRI) explains that the U.S. accounts for approximately two-thirds of Israel's major conventional arms imports, while Germany, the second most important supplier of weapons, contributes nearly 30% (Gritten, 2024).

As the third-largest exporter of arms to Israel, Italy plays a significant role in supporting a state that is committing a genocide. Italian Prime Minister Giorgia Meloni publicly claimed that Italy did not send any arms to Israel after October 7, 2023. She also praised her government's approach as being more rigorous than those of other nations. However, facts disprove these claims. Despite the International Court of Justice's January 26, 2024 order, which found South Africa's genocide allegations to be plausible and required Israel to prevent violations of the Genocide Convention, the Italian government continued to deliver military equipment to Israel under previously approved licenses, even after October 7, 2023 (Reuters, 2024b).

The United Kingdom is also a contributor to Israel's military operations, with arms export licenses valued at €167 million (Akkerman, 2024). However, the United Kingdom is more than just a weapons supplier. Together with the United States, it is also a partner directly involved in carrying out military operations. According to Al Jazeera, of the more than 1,200 reconnaissance and intelligence missions to identify targets in Gaza and southern Lebanon, only 20% were flown by Israeli aircraft. By contrast, 47% were flown by British aircraft and 33% were flown by U.S. aircraft. Most of the bases used by these aircraft were in Germany, Greece, Cyprus, and Italy. The investigation confirmed that these missions were not merely surveillance. These operations facilitated attacks on residential buildings, hospitals, schools, and other civilian infrastructure (Brawn, 2024).

The West's unconditional support for Israel is also evident in its voting patterns regarding ceasefire resolutions in the United Nations Security Council and General Assembly. For decades, media outlets based in Washington, London, and Brussels have portrayed the occupation of Gaza as a complex and divisive issue. However, when analyzing global support and solidarity with Palestine, the reality tells a very simple story: almost the entire world stands united on one side, with Palestine, while the United States, and a few other countries stand with Israel.

In December 2023, the 193-member United Nations General Assembly (UNGA) overwhelmingly supported a resolution calling for a ceasefire in Gaza. Only ten nations voted against it: in addition to Israel and the United States, Nauru (a Pacific island nation with a population of just over 11,000) and seven others (Masih, 2023). In December 2024, the Assembly passed another resolution calling for an immediate, unconditional, and permanent ceasefire. The resolution passed with 158 votes in favor, while the same familiar minority, including the U.S., Israel, Nauru, the Federated States of Micronesia, and Tonga, voted against it (Al Jazeera, 2024c).

Why have Western nations supported Israel's onslaught so blindly? Why risk facing trial at a Nuremberg-style tribunal? How could they allow their history to be stained with the blood of Gaza?

There may be at least two primary explanations: one geopolitical and the other cultural.

Geopolitically, the "special relationship" between Israel and the West has long functioned as a mutually beneficial alliance. The United States and Europe depend on Israel to enforce their strategic interests in the Middle East, and to maintain regional control, surveillance, and deterrence. In return, Israel depends on Western support to pursue its expansionist goals with impunity, shielded from international consequences.

However, there is more to this alliance than geopolitics. A deeper, cultural logic is also at work, one rooted in the legacy of cultural supremacy and a long-standing belief that Western civilization is an inherently superior form of societal organization. This worldview often frames some non-Western societies as backward and uncivilized. Within this cultural and emotional hierarchy, both in Israel and the West, the suffering of Palestinians is easier to dismiss, their humanity easier to deny, and their annihilation easier to justify.

2   Western Civi*lie*sation

The repeated use of the term "the West" may raise concerns among critical readers, especially in academic circles. What does it really mean, anyway?

The concept is highly contested. Countless historians and political scientists have used it in the context of international relations and global power dynamics, particularly with reference to countries such as the United States, European nations, Japan, Canada, Australia, New Zealand, and Israel. Nevertheless, the idea remains controversial. It is not even certain where exactly the West is located. This is not fixed, neither geopolitically nor geographically. According to Subhadra Das, author of *Uncivilised: Ten Lies That Made the West*, the West is "where white people are" (Das, 2025, p. XV). However, this definition is ambiguous, since intercultural exchange has significantly transformed the ethnic and demographic composition of Western societies, especially in the last century.

Moreover, the term "West" is problematic because it is not a monolithic, homogeneous geopolitical category. It comprises a wide array of diverse nations and individuals with very different political positions. During a presentation at *King's College London*, I asked Ilan Pappé, professor at Exeter University, about the West's emotional indifference to the genocide in Gaza. He eloquently replied, "the West is also you and me."

Criticism about the use of the term "the West" also arose during a talk I gave, in Istanbul, when a member of the audience suggested I drop it entirely. A colleague at the *University of Venice* offered a similar critique after briefly reading the main contents of this book, remarking that "the West is an inflated term."

If the concept of "Western civilization" is fraught with conceptual issues, the idea of defining the world's geopolitical borders based on the division between the West and the Rest is an even bigger problem. A significant source of criticism toward this categorization relates to the political background of the very authors who helped pioneer it. Perhaps, the most notable example is political scientist and historian Samuel Huntington. In *The Clash of Civilizations and the Remaking of World Order*, the author popularized the idea of framing global conflicts and cultural divides in terms of Western and non-Western powers. Huntington argued that the West's success over other civilizations is due to a unique combination of factors, including superior military power and economic and technological advances, but also political organization and cultural values of creativity, innovation, individualism, pluralism, and human rights (Huntington, 1996).

Another prominent academic who has popularized the idea of the "West and the Rest" is Niall Ferguson. In his book *Civilization: The West and the Rest*, Ferguson examines the factors behind the rise of Western nations to global dominance, particularly those in Europe and North America. Like Huntington, Ferguson attributes the West's success to a combination of technological innovation, political and economic institutions, and cultural values. He contrasts these characteristics with those of other civilizations, arguing that they did not develop similar advantages to the same degree. Although other regions faced different historical and structural challenges, Ferguson believes that the characteristics that gave the West a comparative advantage were innovation, individualism, and the rule of law (Ferguson, 2011).

Despite the limitations of the term "West," categorizing geopolitical blocs as Western or non-Western is simple and easy to understand. Furthermore, the West's role in the Gaza genocide is undeniable. I agree with Ilan Pappé that "the West is also you and me." However, for almost two years, neither he, nor I, nor the nations that opposed Israel from the outset, such as Ireland and Spain, have been powerful enough to exert sufficient pressure to stop the genocide.

Moreover, adopting the term "the West" does not necessarily imply an endorsement of the political positions of Huntington and Ferguson, as some academics seem to suggest. In fact, it can be the exact opposite. The limitations of their arguments deserve scrutiny. Both authors tend to promote and reinforce a favorable image of Western identity and dominance, while downplaying, or outright omitting, uncomfortable historical realities, such as systemic racism, colonization, and exploitation. Their failure to account for the complex and often violent mechanisms through which power and wealth have been produced, exercised, and maintained is especially problematic. In their books, these dynamics are, at best, obscured and, at worst, treated as natural or inevitable features of global development.

One of the most glaring omissions from mainstream explanations of the West's geopolitical and economic success is the foundational role of African slavery and the massive extraction of natural resources such as gold and silver from Latin America. These two factors were crucial to the initial accumulation of wealth that fueled Europe's Industrial Revolution. Yet some commentators deny these connections entirely. For example, Milton Friedman, one of the most influential neoliberal economists of the 20th century, famously claimed that "Britain ... did not have slavery" and dismissed the idea that "the enormous increase of world wealth in the free countries of the West was due to slavery" (Freedom Channel, 2007).

Another compelling argument for deconstructing mainstream historical interpretations of global geopolitical affairs is the role that colonization and various forms of subjugation have played in shaping the world as it is today. Colonization can take the form of direct territorial invasion or more subtle forms of external control, often supported by systems of economic exploitation

(Kohn & Reddy, 2024). While power struggles, conquest, and oppression have been a part of human history everywhere, Western colonization is unique in that it subjected two-thirds of the globe to invasion, territorial dispossession, and the destruction of indigenous economic and political systems.

The major colonial empires, Britain, France, Spain, Portugal, the Netherlands, Germany, Belgium, and Italy were all Western powers. Their expansion reshaped entire continents. The nations most subjected to this domination include those in Africa, Latin America, Southeast Asia, and the Middle East. Furthermore, settler colonialism, an enduring and dispossessive form of colonization, established Western outposts in regions previously inhabited by indigenous peoples. Australia, Canada, and New Zealand were colonized and largely cleansed of their native populations, which were replaced by Europeans. These regions were transformed into extensions of the West.

As European powers gradually declined, the United States, a settler colonial state born of European expansion, assumed the mantle of global hegemony. Instead of breaking with colonial legacies, the U.S. perpetuated them, becoming the leading power in an era of economic control and imperialistic intervention. From Vietnam to Iraq and from Latin America to the Middle East, American foreign policy has often mirrored the logic of European imperial domination, employing not only direct force, but also softer strategies of power such as economic coercion, cultural hegemony, and regime change.

Israel occupies a unique place in this geopolitical landscape. While geographically located in the Middle East, it is widely regarded as part of the political and cultural construct of "the West." This classification is based on strategic alignment, shared military interests, and ideological proximity to Western powers, particularly the United States.

2.1 *Genocidal Legacies*

It was the fall of 1997, and I had just moved to Los Angeles to begin a master's program in public health at the *University of California Los Angeles* (UCLA). It was a period of intense self-reflection and reading, driven by a growing sense of dissatisfaction, even frustration, with the mainstream university curriculum. While the courses I attended covered the determinants of diseases and mortality, they rarely explored the structural, historical, political, and economic forces that underlie health outcomes, especially in developing countries. Something, I felt, was profoundly missing from the analysis offered by most global health faculties.

Two books changed the way I viewed global health and politics, my culture and Western civilization: *How the Other Half Dies: The Real Causes of World Hunger* by Susan George (1976) and *Year 501: The Conquest Continues* by

Noam Chomsky (1993). Reading them was a humbling experience. Like many students raised and educated in Western societies, I had initially approached global health through a culture-centric lens, assuming that Western policies, humanitarian interventions, and foreign aid were primarily altruistic efforts aimed at helping the world's most disadvantaged. Susan George's and Noam Chomsky's books first shattered this naïve illusion.

These books, among many others, merit recognition for challenging some of the most deeply entrenched myths propagated by mainstream media and Western intellectual culture. Susan George dismantled the prevailing notion of Western countries as benevolent Samaritans. She showed that the real causes of world hunger were not the reasons commonly heard on mainstream television programs or read in newspapers such as overpopulation, laziness, lack of education, or cultural backwardness in poor countries. Instead, these causes can be found in the systemic power imbalances, inequality, and an international financial order designed to protect the interests of the wealthiest sectors of society and corporations.

Though many other works before and since have demonstrated that most famines are man-made, George's book served as my baptism into the critical political economy of global health. I still remember a lonely night in the UCLA Biomedical Library, lost in the pages of *How the Other Half Dies*. The book revealed that world hunger was rooted in the systematic injustices perpetrated by Western nations, international financial institutions, and transnational corporations. Even financial aid often exacerbated the situation. I was shocked, saddened, and profoundly guilty. I had naively assumed that Western aid to the developing world could be part of the solution. In reality, it was part of the problem. Since then, my understanding of power politics has changed, and the world has never looked the same.

Reading Susan George marked the beginning of a personal journey to uncover how my cultural biases had shaped my view of history and international relations. But it was Noam Chomsky's *Year 501: The Conquest Continues* that proved to be a decisive turning point in furthering my consciousness transformation. I read the book twice. The second time while working as a short-term consultant for the World Bank in Washington, D.C., a setting that made the book's message even more powerful. Years later, I published an academic paper reviewing the harmful health effects of World Bank and International Monetary Fund (IMF) economic policies on women and children living in Sub-Saharan Africa (De Vogli & Birbeck, 2005). This seemed like a very promising way to secure a stable, future career at the World Bank, right? After all, what organization wouldn't want to hire a researcher who exposed the lethal policies of his own employer?

My personal story is not important, but the subject is. More than any other publication I read at the time, Chomsky's book made me aware of two pivotal historical moments that fueled the global rise of inequality: the European colonial expansion, which began with the conquest of Latin America in 1492 and the start of the transatlantic African slave trade from Hispaniola in 1501. Chomsky writes that the exploitation of the so-called New World "unleashed vast demographic catastrophes unparalleled in history: the virtual destruction of the indigenous populations of the Western Hemisphere and the devastation of Africa as the slave trade expanded rapidly to meet the needs of the conquerors" (Chomsky, 1993, pp. 5, 501).

In his powerful critique, Chomsky draws on the work of one of the most frequently cited, but often selectively read, figures in neoliberal and conservative thought: Adam Smith. The moral philosopher and Scottish economist defined the discovery of America and the passage to the Orient through Cape of Good Hope as the two most important events in human history. Smith observed that, "the discovery of America certainly made a most essential contribution to the state of Europe ... opening a new and inexhaustible market" that brought "real revenue and wealth." But Smith also offered a rare moral reflection that is often overlooked by his neoliberal admirers: "The cruel injustice of the Europeans has made an event, which ought to have been beneficial to all, ruinous and destructive to several of those unfortunate countries" (Smith, 1776, pp. 343, 26).

The capture of the Inca emperor Atahualpa by Francisco Pizarro and of the Aztec emperor Montezuma by Hernán Cortés marked the beginning of what can be called the "Nakba for the Native Americans." While it is true that the indigenous peoples of the Americas were not pacifists, they initially approached the Europeans without aggressive intentions. As Adam Smith noted, "far from having ever injured the people of Europe, [the Native Americans] had received the first adventurers with every mark of kindness and hospitality" (Smith, 1776, p. 190). It proved to be a tragic mistake.

Christopher Columbus, describing the first Indigenous people he encountered, remarked: "they are so naïve and so free with their possessions that no one who has not witnessed them would believe it. When you ask for something they have, they never say no. To the contrary, they offer to share with anyone." He then added: "they would make fine servants ... with 50 men we could subjugate them all and make them do whatever we want" (Zinn, 1980, pp. 3–4).

The conquest and occupation of America has long been celebrated as a defining achievement of Western civilization. This narrative has been reinforced by Hollywood's enduring portrayal of Native populations as villains, which casts the original victims as aggressors in a striking reversal of historical roles. Some authors, however, have challenged these historical distortions.

A Short Account of the Destruction of the Indies by Bartolomé de las Casas is perhaps the first eye-opening description of the adventures of the West in Latin America (de Las Casas, 1999). After a moral and spiritual awakening, de las Casas dedicated most of his life to campaigning against what he deemed as the unparalleled brutality of Spanish soldiers against Native Americans. As he wrote, "I saw here cruelty on a scale no living being has ever seen or expects to see" (Sullivan, 1995, p. 146).

In *The American Holocaust: The Conquest of the New World* David Stannard offers a searing and meticulously researched account of the catastrophic impact of European colonization in the Americas (Stannard, 1992). Stannard documents in harrowing detail the widespread genocide, cultural destruction, and systematic oppression inflicted on indigenous communities. Like de las Casas, he argued that the scale and brutality of these atrocities rival or surpass those of the most notorious genocides in modern history.

The popularity of these types of exposés and protests by activists committed to decolonizing history has made the celebration of the "discovery of America" in the West as a source of global historical reckoning. The mass slaughter and exploitation of Native Americans is indeed one of the most devastating and extensive genocides in recorded history. When Columbus arrived in the New World in 1492, the continent was home to an estimated 60 million indigenous people. By 1600, approximately 90% of this population had been wiped out, through direct violence, forced labor, and the structural deprivation imposed by colonial systems (Koch et al., 2019).

However, historical narratives that obscured the real causes of this genocide are not in short supply. Influential works by Western authors, such as UCLA professor Jared Diamond's *Guns, Germs, and Steel*, have attributed the demographic collapse in the Americas primarily to infectious diseases (Diamond, 1997). Yet this explanation is deeply misleading as it ignores the importance of the settler colonial determinants of health (Wispelwey, Tanous, et al., 2023). As social epidemiologists have long emphasized, infectious diseases do not kill in a vacuum. Epidemics are driven by social, economic and political conditions that affect health (Li et al., 2024) including poverty, malnutrition, overcrowding, forced relocations, poor sanitation, lack of access to clean water, and the psychosocial toll of systemic oppression and exploitation.

Following the catastrophic decline of Indigenous populations in South America, whose forced labor was instrumental in enriching European empires, Western powers turned to the African continent to sustain their extractive economies. Between the 16th and 19th centuries, an estimated 12 million Africans were forcibly transported across the Atlantic to work as slaves on plantations. Between 1750 and 1830 alone, approximately 15% of enslaved Africans died during the brutal Middle Passage with countless more perishing during capture and transport long before reaching the ships (Mancke & Shammas, 2005, Galeano, 1973, pp. 11–12).

Europe's atrocities in Africa did not end with the transatlantic slave trade, however. They continued for centuries through the violent machinery of colonialism. One of the most egregious examples occurred between 1885 and 1908, when King Leopold ii of Belgium transformed the Congo into a vast, ruthless labor camp. Colonial forces under Leopold's rule committed atrocities on an unimaginable scale, including mutilations, executions, and mass starvation, leading to the deaths of an estimated ten million Congolese (Montero & Lowes, 2019).

France, too, bears a bloody colonial legacy. In Algeria, it maintained a brutal grip until 1962, relinquishing the territory only after an eight-year war for independence. The occupation, repression and massacres claimed the lives of up to one million Algerians, and historical records detail the systematic use of torture, as well as the forcible relocation of more than two million people in "regroupment camps" (Horne, 2011).

Although Italy is often romanticized in Western culture with slogans such as *Italiani brava gente*, the country was also responsible for horrific crimes in Africa. In Ethiopia, Eritrea, and Libya, Italian forces operated concentration camps where hundreds of thousands of civilians were imprisoned and starved. These crimes were documented in the BBC 1989 documentary *Fascist Legacy*, (BBC, 1989) the rights of which were reportedly purchased by the Italian public broadcaster *Radio Televisione Italiana* (RAI) two years later. The documentary was not aired for several years, and it was not until 2004 that the private Italian channel *La7* broadcast some excerpts of it. Even today, the documentary remains largely unknown in Italy.

Germany's colonial crimes, though often overshadowed by the Holocaust, also left deep scars on the African continent. From 1904 to 1908, German forces committed the first genocide of the 20th century in present-day Namibia, exterminating around 100,000 Herero and Nama people through mass killings, starvation, and death camps. And while the Holocaust remains a uniquely systematic and industrialized genocide, with six million Jews exterminated under Nazi racial ideology, it was not limited to Jewish victims. The Nazis also targeted millions deemed "sub humans," including Russians, Poles, Roma, the disabled, Serbs, Slovenes, and homosexuals.

The brutalities of British colonialism around the world too have been overlooked or downplayed in mainstream historical narratives. However, there are some notable exceptions. Using mortality data, scholars Dylan Sullivan and Jason Hickel have directly challenged the sanitized view of British imperialism as a relatively benign form of domination. They estimated that approximately 50 million excess deaths can be attributed to British colonial policies between 1891 and 1920 (Sullivan & Hickel, 2023).

Engineered famines were a hallmark of British imperial policy, not isolated incidents. They did not just occur in India (Mallik, 2023) but also in Ireland, South Africa and Kenya. The early 20th century saw the British Empire commit numerous additional atrocities. One of the most notorious was the development of some of the first concentration camps during the Second Boer War in South Africa, where thousands of civilians, mostly women and children, died due to malnutrition, disease, and neglect (Laband, 2015).

Similarly violent logics underlay European settler colonial projects in Canada, (Starblanket, 2018) Australia, (Allam & Evershed, 2025) and New Zealand (Brett, 2015). These colonial activities were characterized by the systematic dispossession of Indigenous lands, the forced assimilation of Indigenous children through residential school systems, and, in many cases, outright extermination. Massacres, forced relocations, and state-sanctioned cultural erasure were carried out to eliminate Indigenous resistance and consolidate settler control.

The United States also bears responsibility for the genocide of Indigenous peoples in North America. This ethnic cleansing unfolded over centuries through a combination of military campaigns, forced removals, broken treaties, and cultural annihilation. One infamous policy was the Indian Removal Act of 1830, which resulted in the *Trail of Tears*, a series of forced marches during which thousands of Native Americans died while being relocated to lands west of the Mississippi River. Beyond physical displacement, settler colonial expansion was accompanied by massacres, such as those at *Sand Creek*. It was also marked by the establishment of boarding schools designed to forcibly assimilate Native children and strip them of their language, culture, and identity. These policies were not accidents of war or the unfortunate byproducts of expansion, but rather deliberate strategies aimed at eliminating the Indigenous presence to make way for settler domination (Dunbar-Ortiz, 2023).

2.2   *Year 533: the Conquest Continues*

Mainstream media pundits often reject the notion that Western civilization has played a predominantly predatory role in history. They argue that war and conquest are inherent features of all empires, not unique to the West. Moreover, they emphasize that modern liberal democracies in Europe and in the United States have championed values such as freedom and human rights, principles often lacking in other regions and cultures. They accuse those who try to decolonize history of perpetuating the same bias they are trying to confront, only in opposition.

It is true that the West has not held a monopoly on genocides, invasions, or conquest. A useful lens through which to assess the contributions of various nations and empires to historical atrocities is the scale of mortality events throughout history. While Europe bears responsibility for some of the deadliest historical events such as World Wars I and II, the genocide of Native Americans, and the Atlantic slave trade, some of the highest annual death tolls ever recorded in human history stem from events caused by non-Western nations including China and Russia (Armstrong, 2019).

In defense of the West, mainstream opinion makers often argue that invasions, and occupations carried out by European powers belong largely to a colonial and imperial past. According to this view, such actions are relics of a bygone era, and Western nations have since evolved, promoting civic values and respect for international law. Moreover, the creation of the *League of Nations*, and later the *United Nations*, is seen as evidence of this progress, reflecting a renunciation by Western democracies of wars of expansion and a focus on defense and aid to nations victimized by foreign aggression.

Despite the deadly legacy of the past, the idea that the West is made up of peace-loving, civilized nations that are committed to protecting universal values of freedom and human rights is widely shared in some mainstream media outlets. This argument is often accompanied by the notion that the major threat to world peace comes from nations such as Russia or the "axis of evil" Iran and North Korea.

This view ignores a crucial reality: some of the world's most intractable problems arise because of Western actions, not despite them. The narrative of Western civilization as "the good guys" fighting evil abroad crumbles under even the slightest scrutiny of historical evidence. Over the past century, the United States, the most powerful member of NATO, showed an unparalleled level of foreign aggressiveness. From its inception, U.S. expansionist ambitions were already evident in the annexation of California, Texas, Arizona, New Mexico, Nevada, Utah, and parts of Colorado and Wyoming. Gore Vidal famously referred to these states as the "Occupied Territories of Mexico."

Since the end of World War II, the United States has escalated its military and geopolitical presence around the world even further. Far from just protecting human rights, the United States has become deeply involved in global domination by establishing an extensive network of military bases and by engaging in hundreds of military actions, ranging from large-scale wars and covert operations to drone strikes and regime-change campaigns. Most of these interventions were not about countering foreign threats. Instead, they were driven by strategic, economic, and ideological interests.

The number of victims attributable to US imperialism is difficult to quantify. A prime example is the conflicts that followed September 11, 2001. According to the *Watson Institute* at *Brown University*, which estimated the number of deaths caused by wars since 9/11, there have been 4.5 million casualties from the direct and indirect impacts of war in countries such as Iraq, Afghanistan, Pakistan, Syria, Yemen, and Somalia (Savell, 2023). While the United States is not solely responsible for these deaths, it has played a pivotal role in all these conflicts and their impact on mortality.

2.3   *Superiority Complex*

Western domination has not been advanced only by military force, but also through "soft power" and cultural influence. The hegemonic reach of European culture, both within and beyond its borders, has been sustained by a deeply rooted belief in the superiority of Western civilization. This ideology accompanied colonization with widespread dehumanization, subjugation, and the systematic erasure of other cultures.

In *WEIRDest People in the World: How the West Became Psychologically Peculiar and Particularly Prosperous*, Joseph Henrich explores how Western, educated, industrialized, rich, and democratic societies have come to represent psychological outliers on a global scale. According to Henrich, a large proportion of people in Western democracies often mistakenly apply their social norms as if they were generalizable to all other cultures in the rest of the world (Henrich, 2020b). Furthermore, a significant proportion of individuals and intellectuals within these societies tend to view Western civilization as a pinnacle of cultural and societal development, sometimes in contrast to other nations perceived as less advanced, and even less civilized.

This sense of cultural supremacy has deep historical roots. Many leading Western scientists, philosophers, and political leaders believed Europe's geopolitical dominance is a natural consequence of biological superiority. One of the most pernicious examples is Francis Galton's *Eugenics Theory*, rooted in race science, which promoted the reproduction of the *fit* while discouraging that of the *unfit* (Gray, 2016). Eugenics found intellectual kinship with Herbert Spencer's *Social Darwinism*, which framed war, famine, and even death as necessary purifying forces for human advancement. In *Social Statics*, a book Spencer wrote while serving as a sub-editor at *The Economist*, he argued that societies naturally expel the "unwholesome" and that it is ultimately better for the "imperfect" to not survive (Spencer, 1995).

Western philosophy too contains strong assertions of superiority over other cultures. In his 1689 work, *Two Treatises of Government*, John Locke implied a hierarchy of civilizations and denied the legitimacy of indigenous peoples' land rights, by claiming that land in America not cultivated according to European standards could be rightfully seized (Locke, 2016). In a footnote to his essay *Of National Characters*, written in 1777, Hume expressed his belief that he was "apt to suspect the negroes to be naturally inferior to the whites" (Hume, 1987). Friedrich Hegel, in his 1821 book *Philosophy of History*, explained that "The Negro … exhibits the natural man in his completely wild and untamed state … there is nothing harmonious with humanity to be found in this type of character" (Hegel 2001, p. 111). He also wrote that, "Among the Negroes moral sentiments are quite weak, or, more strictly speaking, non-existent" (Hegel 2001, pp. 113–114). Finally, he wrote that, "Slavery is … a phase of education—a mode of becoming participant in a higher morality" (Hegel 2001, p. 117).

Extreme views about superior and inferior races, as well as the dehumanization of certain populations, are usually associated with authoritarian regimes, such as Nazism and fascism or brutal dictatorships. However, less attention has been paid to the extreme views about race held by some of the most beloved or influential political leaders in Western democracies. According to Theodore Roosevelt, "The most ultimately righteous of all wars is a war with savages, though it is apt to be also the most terrible and inhuman. The rude, fierce settler who drives the savage from the land lays all civilized mankind under a debt to him." He continued: "it is of incalculable importance that America, Australia, and Siberia should pass out of the hands of their red, black, and yellow aboriginal owners, and become the heritage of the dominant world races" (Roosevelt, 1889, pp. 45–46). Racial supremacist remarks were also typical of another prominent figure of Western civilization, Winston Churchill. During an Iraqi revolt against the British Empire, Churchill complained: "I cannot understand this squeamishness about the use of gas." He continued, "I am strongly in favor of using poisoned gas against uncivilized tribes." In reference to a disastrous famine in the north-eastern region of Bengal, India, then still a British possession, when at least three million people are believed to have died, Churchill blamed the Indians complaining that they "breed like rabbits" (Heyden, 2015).

Rather than challenging the notion of Western superiority, a significant part of the scientific community has historically reinforced it. From Frederic Farrar's 19th-century racial hierarchies, claiming that Aryan-speaking individuals are culturally and intellectually superior to African peoples, (Farrar, 2018) to Lewis Terman's publication *Measurement of Intelligence*, (Terman, 1916) there are many examples of theories and concepts claiming the innate superiority of white or Western populations. These ideas persisted into the 20th century in publications such as *The Bell Curve* by Herrnstein and Murray in 1994 arguing that there is a racial hierarchy in intelligence and positioning IQ as a primary determinant of social stratification (Herrnstein & Murray, 1994).

In contrast, some contemporary scholars have challenged reductionist and racially biased narratives with solid data and rigorous statistical analyses. A case in point is a paper published by Harvard epidemiologist Sankaran Subramanian and colleagues that reanalyzed the data from a 1930 study on race and illiteracy, revealing that the disparities in illiteracy between whites and blacks across U.S. states were primarily due to the racial policies of the Jim Crow states, rather than individual characteristics (Subramanian et al., 2009).

Despite progress in reducing ethnocentric bias in research, philosophy, and politics, the ideological underpinnings of Western superiority have not vanished. Rather, they have evolved into more subtle forms, often embedded in the rhetoric and policies of contemporary liberal democracies. For example, former British Prime Minister Boris Johnson described black people as "piccaninnies" with "watermelon smiles," revealing how colonial attitudes continue to surface in mainstream political discourse even today (Bowcott & Jones, 2008).

3   When the Perpetrator Plays the Victim

While Israel and Western nations have followed completely different historical trajectories, they share a mindset shaped by a history of settler colonialism and supremacism. This ideological continuity spanning European and American history and Israeli contemporary policies is clear when comparing the genocidal rhetoric of present Israeli leaders and opinion-makers with that of Western political figures and philosophers of past centuries.

European and American conquerors have often presented the need for more land, more resources, and more control in terms of defense, civilization, or divine destiny. In the case of Israel, this settler-colonial logic is embodied by the political vision of a "Greater Israel," which reflects a drive for territorial expansion similar to historical Western imperialism.

The dehumanizing language, once used to justify the extermination and displacement of indigenous peoples in the Americas, is echoed today in statements by Israeli officials regarding the population of Gaza. In many ways, Palestinians today are in the same position as Native Americans were in the eyes of European settlers: a population that is seen as expendable and whose very presence is perceived as an obstacle to domination.

Throughout history, empires and colonial projects have also often cloaked their ambitions in lofty language to rationalize their brutalities. They have systematically denied their genocide. To maintain a sense of coherence, they relied on intellectual justifications and falsifications of history. As in the

West's own past, myths are now essential to Israel's effort to justify the horrific destruction of Gaza and to preserve a sense of moral integrity and decency.

Settler colonialists have often used the myth that the territories they invaded and colonized were empty lands open to claim. European conquerors adopted this mindset during their "adventures" in Latin and Central America. This perspective is reminiscent of the Israeli Zionist claim to Palestine, describing it as "a land without a people for a people without a land" (Said, 1999, p. 9).

Another enduring myth draws its power from the idea of divine mandate. Throughout history religion has often been used to legitimize mass violence, displacement, and extermination under the guise of spreading "civilization." Today, Israel relies on the myth of *being chosen*. Like the supremacist worldview that once justified European conquest and genocide, this narrative asserts a divinely ordained right to the land, based not on justice or international law, but on the belief that one people is God's favorite.

Another convenient myth used by Israel and often invoked in Western imperialistic narratives is the conflation of annexation and genocide with the "right of self-defense." This justification is totally divorced from its actual legal and moral meaning. The right to self-defense applies to populations who are attacked or occupied, not to an occupier who initiates aggression and annexation. As Norman Finkelstein put it, referring to Israelis in Gaza and East Jerusalem, "They only have one right: the right to … pack up their bags and leave."

An additional colonial justification regards the "inversion of victimhood." In this morally manipulative framework, the occupying power is cast as the victim, while the dispossessed and besieged are portrayed as the aggressors. This self-serving perspective allows powerful states and media to justify atrocities as acts of security or self-defense while denying the oppressed even the right to resist their annihilation. Any act of resistance by the occupied people is labeled as "terrorism."

Crucially, the mental distortions that long prevented Western citizens from reckoning with the colonial roots of their worldview are now a central feature of Israeli society. Despite overwhelming evidence documenting Western and Israeli injustices, the emotional attachment to national identity offsets the cognitive dissonance resulting from the clash between professed values and historical realities. As a result, when Western and Israeli citizens are asked about the moral standing of their nations, national pride and a sanitized historical memory often prevail over factual knowledge of war crimes and imperial violence.

This distorted way of thinking undermines not only awareness and knowledge about key facts concerning historical, geopolitical subjects, but also the very foundation of empathy and emotional sanity. The insensitivity to the suffering of war victims of colonized people is another casualty of colonization. The colonizer mind, trapped in ideological blindness, denies the humanity of the colonized not because of what they do, but because of who they are. What emerges is a psychopathologized set of emotions shaped by colonial mentality, national identity, and supremacist ideology.

4   Emotional Decolonization

Settler colonialism is not limited to territorial conquest, violence, and displacement, but also assimilation, legal erasure, and the systematic confinement of indigenous peoples, often to reservation camps. It is a transformation designed to erase the indigenous presence while appropriating their land, heritage, and even identity. Settler colonial societies often claim indigenous ancestry or symbolic heritage while actively denying the political and material realities of ongoing indigenous existence (Wispelwey et al., 2023).

Settler colonialism is also an epistemic project that occupies and distorts knowledge systems, beliefs, and behaviors. It heavily relies on various forms of internalization and subjugation that reshape how both the colonized and the colonizer see the world. Citizens of settler colonial states internalize narratives designed to justify morally indefensible actions and preserve national pride. Propaganda, myth-making, and selective history fabricate a guilt-free origin story, often based on the literal and symbolic disappearance of Indigenous peoples.

Colonization requires dehumanization. Yet, this process is not one-directional. It affects both the oppressed and the oppressors. Emotional colonization strips away humanity not only from those who suffer violence, but also from those who inflict it. While victims endure trauma, displacement, erasure, perpetrators pay a different price: the erosion of their empathy and the moral desensitization that comes from justifying and committing harm.

Settler colonialism, in particular, exacts a profound cost on compassion. To maintain a system of domination, it requires self-serving rationalizations that obscure the brutal realities of genocide, land theft and structural violence. These narratives foster intellectual parochialism and create deep cognitive-emotional blind spots, especially among those who benefit the most from such systems. The consequence is not just ignorance, but a diminished capacity to feel for others.

Empirical research seems to reflect this emotional toll. Studies have shown that White students, on average, exhibit lower levels of empathy and compassion compared to their peers from other ethnic backgrounds (Espín et al., 2022).

While socioeconomic privilege partly explains this disparity, (Mechanic, 2021) the empathy gap persists across income levels, suggesting more entrenched cultural and psychological patterns, possibly rooted in histories of racial dominance and moral distancing (Tesfaye, 2017).

However, dehumanization is not an inevitable fate; it can be reversed through a transformation of beliefs and identity. Prominent Jewish scholar Rabbi Yaakov Shapiro believes that it is possible to overcome the psychological conditioning of colonization and dehumanization through mental deprogramming. This process allows individuals to regain their humanity, even after experiencing a collective descent into moral numbness. The path forward for Israelis, he contends, involves therapy to recover from the ideological distortion caused by nationalism and the recognition that Zionism is fundamentally anti-Semitic and anti-Jewish (Media Mondo, 2024).

But can we really decolonize our emotions and affects?

Decolonization is not merely a political or legal process aimed at restoring the rights of indigenous peoples. It also requires a profound transformation of attitudes, thoughts and feelings, and the very psychological and political frameworks that legitimize and perpetuate systems of oppression. The manufacture of collective emotions to generate support for political action against another population has long served as an instrument of colonial power, reinforcing domination, dehumanization, and exclusion. In this sense, emotional colonization is as significant as territorial conquest, and the affects of both the oppressed and the oppressors must undergo a process of transformation.

Critical scholars calling for the decolonization of psychology often emphasize structural and political analyses and caution against focusing on dimensions such as empathy. Their concern is that the empathy framework risks personalizing injustice and distracting attention from the systemic roots of oppression and domination. Critical awareness of the colonial formation or suppression of empathy, however, is no less important. In fact, it is essential. A truly decolonized emotional landscape recognizes that moral disengagement and indifference toward distant suffering are not natural or inevitable, but socially constructed. Our capacity for empathy, including who we feel for and whose suffering moves us, has been shaped and constrained by the same forces that have perpetrated colonial domination. These forces explain why we fail to feel for some.

Some of those who call to "decolonize psychology" not only dismiss the role of empathy but also the role of universalism and emphasize the importance of context-specific histories, cultures, and politics. Rejecting universalism altogether, however, risks replicating the very silos that colonialism created. Decolonizing emotions requires that we move beyond narrow, biased social identities and cultural affiliations. It requires that we affirm the transcultural, transnational nature of empathy and compassion, freeing them from the colonial and national boundaries that have long confined them. Our failure to feel for victims of other cultures, our selective empathy, is itself a product of these emotional boundaries.

Scholars who call for the decolonization of psychology often extend their critique to the field of methodology, dismissing science, positivism, and empiricism as instruments of colonial domination. However, as with critiques of empathy and universalism, attacking science is a straw man argument, that is, an attack directed at a misleading target that distracts attention from the real issue. Imperialism has not been fueled by science or facts, but rather by the distortion, manipulation, and selective use of knowledge in the service of power, national interests, and exploitation. Freeing oneself from the mental chains of colonization does not require rejecting science altogether, but rather reclaiming it and critically engaging with knowledge to reveal, rather than conceal, structures of injustice.

Decolonizing emotions means unlearning the borders of compassion that nations and cultures have drawn for us. This process requires a willingness to recognize that every life is equally worthy of grief.

When the Western world learns to shed for Hind Rajab the same tears it wept for Anne Frank, perhaps it will begin to decolonize not only its memory but also its empathy.

4.1 *Pedagogy of the Oppressors*

Paulo Freire's *Pedagogy of the Oppressed* is one of the most influential contributions to critical education ever written. It focuses on the struggle for liberation from systems of oppression through the development of critical consciousness, the ability to identify and confront the social, political, and economic forces that perpetuate domination and inequality (Freire, 1970).

While Freire emphasized the need for transformation in both the oppressor and the oppressed, his pedagogy is grounded in the experiences of the most marginalized people. Yet significant social transformations can also occur when people from nations that oppress other nations rebel against their own leaders. In the context of the West and Israel, a pedagogy of the oppressors, one that confronts the emotional and ideological conditioning of dominant societies, may be even more consequential than merely empowering the oppressed.

Decolonizing the emotions of citizens of nations that perpetuate or support social injustice, settler colonialism, occupation, and genocide requires engaging in a dual process of deconstruction and reconstruction.

The first phase of emotional decolonization (deconstruction) requires dismantling the geographical, cultural, racial, and psychological walls that colonialism has erected around our ability to identify with humanity. Cognitively, it means confronting historical interpretations shaped by propaganda, ideology, and bias in knowledge production. It consists of exposing the selective narratives that feed national supremacy, revisiting recent and historical conflicts with a new pair of critical eyes, and confronting suppressed facts that challenge the sanitized stories nations tell themselves.

The influence of nationalism on our understanding of history is both profound and easily observable, manifesting itself in the blind spots that every nation keeps in its memory. How many British, French, German, and Belgian citizens are aware of their countries' brutal histories in India, Algeria, Namibia, and the Congo, respectively? How many Italians still believe they "civilized" Libya and Ethiopia and how many of them know about the Italian concentration camps in former Yugoslavia? How many Russians can accurately recount the events of the Prague Spring or the Soviet invasion of Afghanistan? How many Chinese citizens know about the history of repression in Tibet? How many Japanese people are taught about the atrocities committed by their army in Manchuria? How many Western citizens know that, before being celebrated as icons of peace and justice, both Gandhi and Mandela were once branded as "terrorists"?

Deconstructing colonized emotions requires a reckoning with the historical deceptions that sustain empires. This is a radical, often disruptive process that confronts not only facts, but also deep-seated emotional tendencies to empathize with our in-group and dismiss out-group narratives. A truly decolonial pedagogy must engage both the mind and the heart, challenging myths, biases, and emotional habits. Such deconstruction involves disrupting the self-serving imperial narratives masquerading as history. It forces people to confront why some war victims are mourned and widely discussed, while others are rendered almost invisible and barely covered by the media.

The second part of emotional decolonization consists of a phase of reconstruction. Deconstructing colonized narratives is not enough. Emotional decolonization is not only about dismantling falsehoods; it is also a constructive psycho-political endeavor. Reconstruction involves learning the hidden histories, counter-narratives, and knowledge that challenge the status quo to build a broader sense of humanity that transcends national, racial, or cultural boundaries. It requires cultivating the emotional capacity to see victims of distant wars, refugees, and marginalized peoples as fully human, deserving of the same grief and solidarity as our own kin. It involves empathizing with distant victims of oppression as if they were close to us.

As Howard Zinn showed in *A People's History of the United States*, seeing history through the eyes of the colonized radically reshapes not only the narrative but also our consciousness and social identity (Zinn, 1980).

Decolonial thinking challenges the intellectual dominance of the West, recognizing that ethnocentric biases have been deposited in our minds through years of schooling, socialization, acculturation, and identity formation. The process of decolonizing emotions therefore involves cultivating openness to a wider range of feelings and perspectives that create space for new ways of imagining empathy, justice, and peace. It requires breaking out of the narrow confines of the "in-group" perspective and selective compassion, cultivating the emotional and intellectual conditions for universal dignity, where every individual is seen as fully human regardless of geography, ethnicity, or history.

This is not just an act of political resistance, but of moral and emotional outreach, designed to develop a new sense of solidarity, a new world where justice is not selective, and where peace is not a privilege of the powerful, but a right for all.

CHAPTER 5

Manufacturing Exclusive Compassion

When covering the Gaza genocide, Western mainstream media seem rooted not in democracy but in hypocrisy.

∴

October 7, 2023, will be remembered as the day of the deadliest mass murder of Jews since the Holocaust. At 6:30 a.m., Hamas, Islamic Jihad, and other armed groups launched a coordinated attack on Israel. The assault began with a barrage of some 1,000 rockets, targeting both military bases and civilian areas throughout Israeli territory.

Then, Hamas commandos breached the Gaza-Israeli border at multiple points. Using motorcycles, pickup trucks, and bulldozers to tear through the fence, they opened the way for larger vehicles. Additional infiltrations occurred by sea, using motorboats, and by air, using motorized paragliders. Several kibbutzim, border towns, and the Nova music festival, a rave gathering of young people, were targeted. This was not only a military operation. It was a massacre of civilians, including women and children.

Some have asked why certain voices, so quick to express solidarity with Palestinians suffering under occupation and genocide, have remained indifferent to the victims of October 7.

The answer is simple: they should not have.

Every innocent victim of violence deserves to be mourned. Those who died on October 7, as well as those who died before and after. On October 6, 5, 4, 3, and all the days before. On October 8, 9, 10, and the days that followed.

1 Israel's 9/11: Hamas's Crimes against Humanity

Some glorified Hamas's horrific atrocities as brave acts of resistance and praised the attack, but the killing and kidnapping of civilians constitute war crimes and may amount to crimes against humanity. While Hamas targeted several Israeli military bases and police stations, and Palestinians certainly have the right to defend themselves against an occupation, *Human Rights*

Watch reported that "Palestinian fighters fired directly at civilians, often at close range, as they tried to flee. The attackers hurled grenades, shot into shelters, and fired rocket-propelled grenades at homes". The report also added that they "set houses on fire, burning and choking people, and forcing out others whom they shot or captured. They took dozens hostage and summarily killed others" (Human Rights Watch, 2024a).

The attacks left many innocent people dead or wounded. According to the Independent International Commission of Inquiry on the Occupied Palestinian Territory, including East Jerusalem, approximately 1,200 people died. Of these casualties, more than 300 were members of the Israeli security forces and about 850 were civilians, including 33 children. Nearly a quarter of the civilians who lost their lives were at the Nova music festival. Approximately 14,970 people were injured and taken to hospitals for treatment. In addition, some 250 civilians (including 36 children, 90 women and elderly individuals) and soldiers were taken hostage with the explicit intention of exchanging them for Palestinians held in Israeli prisons.

Forensic investigations from various sources, including body camera footage of deceased Hamas militants, have revealed how some Israeli civilians died (Tkacik, 2024). There is little doubt that most were killed by Hamas. Some, however, were murdered by the Israeli army because of the so-called *Hannibal Directive*, a doctrine named after a Carthaginian general who poisoned himself rather than be interrogated by his Roman captors. The directive permits the Israeli army to open fire on its own troops if they risk being captured by the enemy.

Relatives of civilians killed in a kibbutz in southern Israel during the October 7 attack have called on the military to further investigate these events (Frankel & Bernstein, 2024). While attributing blame to the Israeli army for the deaths of its own citizens sounds like a conspiracy theory, even the Israeli army chief admitted that an unknown number of civilians and soldiers died from "friendly fire" from Israeli Apache helicopters and tanks (NDTV, 2025). It is still unclear how many of the October 7 casualties were caused by these counterattacks. However, these individuals would certainly not have lost their lives if Hamas had not carried out these attacks.

The 236-page *Human Rights Watch* report entitled *I Can't Erase All the Blood from My Mind* meticulously investigated the sites of the October 7 civilian attacks. Based on interviews with 144 people who witnessed the assaults, victims' family members, first responders, and medical experts, as well as the verification of 280 photographs and videos of the attacks that were circulated on social media, the report details indiscriminate violence against civilians, hostage-taking and other war crimes. These include cruel and other inhuman treatment, mutilation and looting of corpses, using human shields and pillaging and destroying property without military justification (Wille, 2024).

Human Rights Watch also "documented evidence of acts of sexual and gender-based violence perpetrated by fighters, including forced nudity, and posting sexualized images on social media without consent. However, due to the challenges in interviewing survivors or witnesses of rape, the organization was unable to gather verifiable information" (Wille, 2024).

Previously, a *United Nations Office of the Special Representative of the Secretary General on Sexual Violence in Conflict* concluded that "there are reasonable grounds to believe that conflict-related sexual violence occurred during the October 7 attacks in multiple locations across the Gaza periphery, including rape and gang rape, in at least three locations" (United Nations, 2024a). According to Israeli authorities and police, there is video evidence, and photographs of the victims' bodies as well as testimony from militants confirming accounts of sexual assault. Based on such evidence, in November 2024, the ICC issued an arrest warrant for Hamas leader Mohammed Deif on charges of war crimes and crimes against humanity, including rape and sexual assault against captives in Gaza.

However, in an interview published in the *Yedioth Ahronoth* newspaper, Israeli State Attorney Moran Gez lamented that 15 months after the events, Israel had still not identified specific victims against whom an alleged perpetrator of a sexual assault could be legally prosecuted. Gez admitted that it is very difficult to prove these crimes and she observed: "in the end, we do not have any complainants" (Middle East Monitor, 2025b).

Sexual crimes are indeed difficult to investigate, especially in the context of war and conflict.

But there have also been examples of the misreporting of sexual violence, as documented by a piece of the news agency *Associated Press* (Goldenberg & Frankel, 2024). A *New York Times* article entitled *Screams Without Words: Sexual Violence*, which was particularly impactful, and likely contributed to exacerbating the Israeli response at a time when there was already ample evidence of mass killings of innocent civilians in Gaza, was revealed to be misleading and to include false information. More than 50 tenured journalism professors from prestigious universities wrote a letter calling on *The New York Times* to address questions about the accuracy of its own investigation (Wagner, 2024).

In particular, evidence gathered from relatives of Gal Abdush, a woman killed in the attack who became the focus of the *New York Times* article, casts serious doubt on the accuracy of the U.S.-based media outlet's reporting. As Gal's sister stated, "My sister was not raped ... there is no proof that she was raped" (Blumenthal & Maté, 2024). Months later, *The Times* of London also concluded that the hypothesis that Hamas used rape as a weapon on October 7 "does not stand up to scrutiny" (Philp & Weiniger, 2024).

2   "Lied into Genocide"

If the evidence of whether Hamas committed mass rape against civilians on October 7 is still missing, proof of Israel's systematic use of sexual, reproductive, and other forms of gender-based violence since then is indisputable. A report titled "More Than A Human Can Bear" described the deliberate attacks by Israeli security forces on sexual and reproductive health care facilities and the collapsed health care infrastructure in Gaza. The document also reported on the sharp increase in sexual and gender-based violence, including rape and other forms of sexual violence, perpetrated by members of the Israeli security forces and settlers throughout the Occupied Palestinian Territory (OHCHR, 2025b).

In the aftermath of October 7, while Israel had already inflicted widespread death and destruction in Gaza, much of the Western mainstream media remained fixated on the events of that day. Some reports were grounded in verified facts and serious investigations, but others were based on fabricated information. False reports included shocking claims such as a baby allegedly placed in an oven, a pregnant woman having her fetus stabbed and then shot, and the beheading of infants. None of these stories have been supported by evidence (Goldenberg & Frankel, 2024; Maad et al., 2024).

A commander and a volunteer from ZAKA, an Israeli community ambulance and rescue organization, made a series of statements about acts of sexual violence and rape. Photographs allegedly showing extreme genital mutilation of women, with nails and knives inserted into the groin, were said to have been provided by ZAKA members, and later used by the UN Special Representative of the Secretary-General on Sexual Violence in Conflict to accuse Hamas. However, none of these images could be independently verified (Goldenberg & Frankel, 2024; Feminist Solidarity Network for Palestine, 2024). In fact, they were revealed to be pure fabrications.

The *New York Times* lamented the "deluge of online propaganda and disinformation larger than anything seen before" in an article addressing various wars, including what the authors described as the "conflict between Israel and Hamas" (Myers & Frenkel, 2023). This is quite correct, but it is a flood of propaganda and disinformation that the *New York Times* itself contributed to spreading. In the aftermath of October 7, the establishment media, with a few exceptions, eagerly recited and repeated Israel's narratives in support of its military operation, rather than investigating the facts and available evidence. Moreover, as Israeli media outlet *Haaretz* noted, evidence of crimes committed by Hamas did not need to be "contaminated by unverified stories disseminated by Israeli search and rescue groups, army officers and even Sara Netanyahu" (Hasson & Rozovsky, 2023).

Indeed, Hamas committed many war crimes, but the Israeli government and some Western media outlets preferred to focus on crimes that were never committed. Why fabricate atrocities that never occurred instead of focusing on those that did?

There are several possible explanations. Some might attribute it to unprofessionalism, confirmation bias, or the emotional chaos unleashed after October 7. Perhaps. But a more disturbing motive is also worth considering. In the context of propaganda, vivid, gruesome, emotionally charged stories serve a strategic purpose; they have a far more powerful influence on public opinion. These fabrications helped justify what followed: the mass killing of Palestinians. War crimes are easier to sell when the enemy is dehumanized, portrayed not just as an opponent, but as a monster.

Data, facts, and verified reports do not evoke the same emotional response. But graphic stories of unspeakable cruelty, repeated ad nauseam, inflame hatred, cloud judgment, and silence dissent. Even Prime Minister Netanyahu seemed to acknowledge he adopted this tactic. While praising the ZAKA volunteers, including the one who spread false stories, he remarked: "We need to buy time by appealing to world leaders and public opinion. You have an important role to play in influencing public opinion, which also influences leaders" (Prime Minister's Office, 2023). The strategy certainly worked. The most powerful political figures and mainstream media outlets in the West echoed Netanyahu's and the ZAKA volunteers' fake news, without even trying to fact-check it.

As journalist Jonathan Cook put it: "We were lied into genocide" (Cook, 2024).

3 "Have You Condemned Hamas?"

Every time a ceasefire is called for, every time the mass killing of Palestinian civilians is exposed, every time people around the world stand in solidarity with Gaza, the same question is hurled like a weapon at anyone who dares to criticize Israel's actions: "But have you condemned Hamas?"

Most of us have. I have. Repeatedly, publicly, and unequivocally.

Yet those who ask this question so persistently have rarely, if ever, condemned Israel's crimes. Have they said a word about the 750,000 Palestinians who were expelled in 1948 during the Nakba, or the catastrophe that turned entire families into refugees overnight? Not once. Did they raise their voices when Gaza became, in the words of *Human Rights Watch*, "an open-air prison" or, as Josep Borrell put it, "an open-air cemetery"? Silence. Did they condemn the massacres of civilians during military operations like *Operation Cast Lead* (2008–2009) or *Operation Protective Edge* (2014), often referred to as "mowing the lawn"? They did not. Have they spoken out against armed settlers, supported, and encouraged by the Israeli state, who terrorize, dispossess, and murder Palestinian civilians in the West Bank? No. Have they condemned the starvation of children, some of whom have already died of hunger? Absolutely not. Have they condemned the destruction of most of Gaza's homes or the repeated forced displacement of 2.2 million people? Never. Have they said anything about the annihilation of Palestinian universities and almost all schools in Gaza? Not even a whisper. Have they condemned the bombing of hospitals and ambulances, and the blockades that have forced doctors to amputate limbs without anesthesia? No. Have they expressed outrage at the killing of medical personnel, aid workers, journalists, and UN staff?

The answer, again, is no. They never uttered a word about these crimes. They have said nothing in defense of the oppressed, but have remained loyal to the perpetrators of genocide. And they cannot even tolerate expressions of solidarity with the right of Palestinians to exist, and resist.

4   The Engineering of Selective Indignation

Social psychologists have long studied why we empathize more with members in our in-groups than with those in out-groups. However, far less attention has been given to how mass media shape and reinforce these emotional biases. This oversight becomes particularly significant when considering the disparity in public compassion for victims of war across different regions. Mass media play a pivotal role in shaping public understanding of global events. Through choices in framing, narrative vividness, and the attribution or omission of blame, media outlets can either challenge or reinforce existing stereotypes and prejudices. These editorial decisions deeply influence the level of empathy extended toward distant others.

World-renowned psychologist Albert Bandura studied the main mechanisms of moral disengagement including self-serving justifications, minimizing other people's suffering, victim blaming and using sanitizing language. The latter is particularly important in the context of mass media. As Bandura put it: "Language shapes thought patterns on which actions are based. Activities can take on very different appearances depending on what they are called" (Bandura, 1999). Euphemistic or inflammatory language, for example, can shape reality by downplaying some forms of aggression while magnifying others.

One example of how our emotions have been directed, or misdirected, is the stark contrast in the media's portrayal of war crimes in Ukraine and Gaza. Since February 24, 2022, the suffering of Ukrainian civilians has dominated the front pages of Western mainstream media. Unlike in previous conflicts, such as the U.S. invasion of Iraq, Russian war crimes have been extensively documented and reported. Powerful, emotionally charged images have filled the news: bodies exhumed from mass graves, mothers weeping over their sons' corpses, civilians digging through rubble for loved ones, refugees fleeing bombed-out cities, and wounded children lying in hospital beds. These were truly heartbreaking images that deserved our full compassion, and outrage. The language used to describe these horrors was graphic and unequivocal in assigning responsibility to Russian forces.

In contrast, despite a far higher number of civilian casualties, and a far higher civilian-to-combatant ratio, coverage of Gaza has been far more restrained. The tone has often been more neutral, abstract, and evasive, with significantly less graphic imagery and fewer or no direct accusations. Media descriptions of Israel's actions in Gaza, widely condemned by human rights organizations as genocide, often relied on vague, indirect language, thereby reducing clarity and accountability.

One of the clearest examples of this double standard is the repeated use of the passive voice when reporting on Israeli airstrikes in "safe zones," which are areas designated by the Israeli army as places where civilians are instructed to go to avoid harm. For example, a *Financial Times* headline that read: "Dozens killed and wounded after explosions at Gaza 'safe zone' camp. Local authorities blame Israeli air attack, while Israel says it was targeting Hamas compound" (Zilber, 2024). This framing clearly disregards Israel's role as the direct perpetrator of violence against civilians. A more accurate and direct headline might have read: "Israel kills and wounds Palestinian civilians in designated 'safe zone.'" Instead, the actual wording seems to soften the brutality by suggesting coincidence or ambiguity where none exists.

In November 2024, over 230 media professionals, including more than 100 BBC employees, signed an open letter to BBC Director General Tim Davie. The letter accused the organization of exhibiting a significant bias in its Gaza coverage. The professionals criticized the network for using "dehumanizing and misleading headlines," pointing to its coverage of clear cases of killings by Israeli soldiers, such as the assassination of Hind Rajab. Despite well-documented evidence, the BBC headline deliberately avoided identifying those responsible for her death (Stavrou, 2024).

In a compelling editorial published by *The Independent*, Karishma Patel, a former BBC newsreader explained her decision to leave the media organization. Denouncing the distorted coverage and lack of explicit accountability in reporting the killing of five-year old Hind Rajab, Patel concluded: "The BBC failed Hind" (Patel, 2025).

A media analysis of *CNN* and *MSNBC* revealed a stark imbalance in coverage between the Russia-Ukraine war and Israel's military campaign in Gaza. In the first 100 days after Russia's invasion of Ukraine, both networks aired twice as much coverage of civilian suffering in Ukraine as they did of civilian suffering in Gaza, even though Israel's bombing and siege of Gaza resulted in 38 times more child deaths over a comparable period. The disparity was not only quantitative but also qualitative.

Emotive words such as "brutal," "massacre," "slaughter," "barbaric," and "savage" were used overwhelmingly to describe violence against Ukrainians and Israelis, but rarely, if ever, to characterize the mass killing of Palestinians (Johnson & Ali, 2024). The authors explained that the purpose of the analysis was not to argue that the media should have reduced its coverage of the Ukrainian and Israeli tragedies, but rather that the media must have raised its coverage of Gaza's suffering to a level comparable to Ukraine's and Israel's, with the same sense of urgency and moral clarity.

Like the *BBC*, *CNN* has also faced internal criticism for apparent bias. According to six employees and more than a dozen internal memos, the network's editorial policies have effectively censored Palestinian perspectives, prompting a backlash within the newsroom (McGreal, 2024a).

A *Pew Research Center* survey revealed that the U.S. media had misinformed citizens about important international events. Despite the dramatic disparity in casualties (Palestinian deaths outnumbered Israeli by about 26 to 1) about half of the Americans surveyed could not identify which side had suffered more. This confusion reflects not just public apathy, but also how the media shapes, filters, and delivers narratives (Rascius, 2024).

An organization called *Writers Against the War on Gaza/The New York Crimes* conducted a qualitative, comparative analysis of the *New York Times'* coverage of Russian war crimes in Ukraine and Israel's military operations in Gaza. The analysis revealed that the American newspaper consistently and forcefully condemned Russian atrocities in Ukraine, yet it either obfuscated or legitimized and justified Israel's narratives in Gaza. Front-page headlines about Russian crimes included *Horror Grows Over Slaughter in Ukraine* and *Carnage Widens as Ceasefire Talks Falter*, accompanied by graphic photos. In contrast, coverage of the genocide in Gaza used much more neutral titles that justified Israeli actions, such as *Fatal Strike in Dense Area as Israel Hits at Hamas* and *Close Surveillance Led to Israeli Attack on Compound*. The use of certain emotive words was highly selective. *The New York Times* described Israeli deaths as "massacres" 53 times, but only once for Palestinian deaths. Even as the number of Palestinians killed rose to about 15,000, the ratio of the use of "massacre" remained 22 to 1 (Writers Against the War on Gaza, 2024b).

Another analysis by independent media organization *The Intercept*, which examined more than 1,000 articles from *The New York Times, The Washington Post*, and *The Los Angeles Times*, found that emotionally charged terms like "massacre" and "horrible" were almost exclusively reserved for describing Israelis killed by Palestinians (Johnson & Othman, 2024). This did not happen by accident. In a November 2023 memo to the staff, *New York Times* standards editor Susan Wessling wrote: "Words like 'slaughter,' 'massacre,' and 'carnage' often convey more emotion than information" (Writers Against the War on Gaza, 2024a). An internal memo instructed *New York Times* journalists to avoid terms such as "genocide" and "ethnic cleansing" when covering Gaza (Johnson & Othman, 2024). By the editors' own admission, their avoidance of the term "genocide" was a deliberate editorial decision aimed at favoring Israel's narrative.

A content analysis study published in the *Journalism and Mass Communication Quarterly* examined how five popular Western broadcast news channels, BBC News, CNN, *Fox News, Sky News*, and MSNBC, used Instagram in the early stages of what they defined as the "Israel-Gaza war" after October 7. The study found that the channels' coverage emphasized Israeli casualties and framed Israeli violence as self-defense. The results also showed that Palestinian victims were overlooked, while Palestinian violence was not framed as self-defense, but rather as aggression (Hamas Elmasry, 2024).

A report by the *Center for Media Monitoring* also revealed significant bias in the media's coverage of the Gaza genocide. The report analyzed a wide range of data, including 176,627 TV clips from over 13 broadcasters and 25,515 news articles from over 28 British online media websites. The findings highlighted a tendency to ignore the suffering of Gaza's civilians, while portraying Israelis as victims of attacks eleven times more often than Palestinians. The report also showed a substantial level of misrepresentation of Israeli and Palestinian rights. Most TV channels exaggerated and distorted Israel's right to defend itself, overshadowing Palestinian rights of self-determination. Israeli perspectives were mentioned almost three times more often than Palestinian ones. Another recurring theme was the use of euphemisms to describe the Gaza genocide: 76% of online articles framed the conflict as an "Israel-Hamas war," while only 24% mentioned "Palestine/Palestinians." This indicates a lack of context. Finally, the report emphasized the misrepresentation of protests intended to protect the lives of innocent civilians in Gaza. TV channels and newspapers have mischaracterized protests against the genocide in Gaza as anti-Semitic, violent, or pro-Hamas (Centre For Media Monitoring, 2024).

This media bias against Gaza predates October 7, as documented by earlier analyses published in peer-reviewed journals or books on the same subject. Scholarly work has previously documented that *The New York Times* distorted our perception of the "Palestinian struggle" for freedom and self-determination (Jackson, 2024). Another study showed a clear bias in the use of sources, with three out of four mainstream Western newspapers privileging Israeli perspectives over Palestinian ones (Chang & Zeldes, 2006).

4.1 *Hostages and Prisoners*

Another striking example of how the Western mainstream media's manufactured selective compassion is evident in its coverage of the January 2025 ceasefire. The release of Israeli hostages was widely emphasized, while Palestinians freed in exchange were referred to as prisoners, as the BBC called them. Undoubtedly, the long-awaited release of the Israeli hostages deserved applause and front-page headlines. The physical and psychological strain suffered by these individuals after more than a year of imprisonment in Hamas tunnels should have been detailed, as should the concern for their medical conditions (Berg & Moench, 2025). Yet, did it ever occur to BBC journalists that some of the Palestinians released in exchange for the Israeli hostages might have been hostages themselves?

Among those released by Israel in exchange for three young Israeli female soldiers were a 15-year-old boy from East Jerusalem and two 17-year-olds, a boy and a girl (Sky News, 2025). A report by the *UN Special Rapporteur on the Situation of Human Rights in the Palestinian Territory* documented that, prior to October 7, approximately 5,000 Palestinians were imprisoned in Israel, 1,100 of whom were detained without charge or trial. This group included 160 children (OHCHR, 2023a).

After October 7, however, the unlawful practice of detaining individuals without charges became more common. According to a July 2024 UN report, approximately one-third of the 9,500 Palestinians held in Israeli prisons were detained without charge or trial (OHCHR, 2024c). The same UN report found that the individuals referred to as "detainees" by Sky News included women, children, doctors, journalists, and human rights defenders, who were systematically subjected to abuse, torture, and other forms of mistreatment and violations (OHCHR, 2024b). The arbitrary and punitive nature of these arrests and detentions should have provoked some strong reaction in the Western media, but it did not.

In August 2024, the Israeli human rights organization *B'Tselem* released a report titled *Welcome to Hell*, detailing the systematic abuse and inhumane treatment of Palestinians in Israeli custody following the events of October 7. Similar to the UN report, this document is based on the testimony of 55 detainees, including residents of the West Bank and Gaza Strip, as well as Israeli citizens. The report shows that detainees experienced routine physical violence, psychological abuse, and torture at the hands of Israeli authorities. The report also highlighted that detainees were held in overcrowded cells lacking basic hygiene facilities. Numerous accounts detailed the deliberate withholding of medical care from injured or sick detainees. In some cases, this neglect resulted in deaths in custody, some of which were allegedly caused by extreme violence perpetrated by soldiers. *B'Tselem* observed that these abusive practices were not isolated incidents but rather indicative of a systemic policy endorsed by senior officials, including the Israeli government and prison authorities (B'Tselem, 2024).

The report was largely ignored by the Western mainstream media.

The double standard of assigning different weights to human suffering and human dignity became even clearer when describing the release of the three young female IDF soldiers, which was negotiated in exchange for 90 Palestinian women and boys. The Western media's joyful celebration of the soldiers' release could not be compared to the words used to describe the end of the 90 Palestinians' nightmare. The media described all sorts of details about the three Israelis, including their first and last names, hobbies, families, jobs, education, and pets waiting for them (Al Jazeera, 2025a). But one could ask in vain the Western journalists who cheered the exchange of "hostages" for "prisoners" if they had anything to say about the lives, identities, feelings and aspirations of the 69 women and 21 teenage boys (some as young as 12) freed by Israel. Sadly, they did not deserve the same attention. Their release was met with indifference and silence.

Furthermore, while the three female soldiers released by Hamas appeared to be traumatized and physically harmed by the abduction, Israeli doctors certified that "they were in relatively good condition." Pictures of Palestinians released by the Israeli army showed a drastically different health situation and a level of emaciation reminiscent of imprisonment in a concentration camp. Mohammad Sabah is a 20-year-old young man who was imprisoned for 6 years. When he was released from one of Israel's prisons, he was skin and bones, covered with scabies and exhibiting symptoms of severe malnutrition. Coverage in mainstream Western media? Virtually none.

Ibrahim Mohammad Khaleel al-Shaweesh, another Palestinian prisoner from Gaza, was one of the 183 prisoners released as part of the ceasefire agreement between Tel Aviv and Hamas. He reported being subjected to various forms of torture, including electric shocks and the use of dogs. He was blindfolded, handcuffed, and forced to remain on his knees for a month and a half. Arrested on December 10, 2023, al-Shaweesh appeared unrecognizable in photos taken before and after his arrest. The images show the drastic weight loss he suffered during his imprisonment, which left him with a skeletal face and deeply hollowed cheeks, giving him a nightmarish appearance.

Another Palestinian prisoner who appeared in very poor condition after being detained in an Israeli prison is Khalida Jarrar, a human rights activist and an academic at the Birzeit University in Ramallah. She is also a former member of Parliament for the *Popular Front for the Liberation of Palestine* (PFLP). Photos of Khalida in court at the time of her arrest and after her imprisonment are the best proof of the brutalities she suffered. She appeared visibly aged and exhausted; she seemed unrecognizable from the determined and battle-hardened woman she was before her imprisonment. Her homecoming video in al-Bireh, a village near Ramallah, evoked deep emotion among countless activists and concerned individuals, but received no attention from the Western mainstream media.

Khalida spent her time in prison at the infamous Neve Tirza in Ramleh, which is well known among human rights activists for its harsh conditions. She was confined to a tiny 2.5-by-1.5-meter cell. The cramped space contained only one piece of furniture, a concrete bench used as a bed, as well as a dilapidated toilet. In order to breathe properly, she had to lie on the floor with her nose pressed against the crack under the door, as there was no ventilation. Her crime? Nothing. Khalida Jarrar spent six years in prison under administrative detention, without specific charges or trial.

After her release, the first thing she did was to visit the grave of her daughter Suha, who died of cardiac arrest in July 2021. Khalida was denied a temporary release to attend the funeral.

Since October 7, we have repeatedly seen and heard calls to "Free the Hostages" captured and held by Hamas. Anyone with a sense of justice found these demands legitimate, urgent and important. However, some argued that the slogan ought to be revised to "Free All Hostages" to include the approximately 3,000 Palestinians, including women and children, who are being held in Israeli prisons without trial or charge. While Western mainstream media continued to refer to them as prisoners, they were hostages too, like the Israeli civilians held captive in Gaza.

4.2   *"They Look So Much Like Us"*

The evidence showing that Palestinians received far less sympathetic and humanizing media coverage than Ukrainians led some to wonder why. What explains the difference in coverage? Perhaps, the answer was unintentionally provided by *CBS News* senior foreign correspondent Charlie D'Agata, who remarked that Ukraine "is not like Iraq or Afghanistan, which have been in conflict for decades. It's a relatively civilized, relatively European—I have to choose those words carefully—city where you wouldn't expect this to happen." D'Agata likely assumed that conflict was to be expected in Iraq and Afghanistan, but not in Ukraine. Yet, one can't help but wonder what his comments might sound like if he was not careful with his language (Nugent, 2022).

In an interview with the *BBC*, a former Ukrainian deputy prosecutor general remarked, "it's very emotional for me because I see European people with blue eyes and blond hair being killed every day," when referring to the refugees caused by the Russian invasion after February 24, 2022. The *BBC* anchor responded simply: "I understand and respect the emotion" (Ellison & Andrews, 2022).

Similarly, on France's news broadcasting television and radio network *BFM TV*, journalist Philippe Corbé stated: "We are not talking about Syrians fleeing the bombing of the Syrian regime, backed by Putin. We're talking about Europeans leaving in cars that look like ours to save their lives" (Ali, 2022). An *ITV* journalist reporting from Poland explained, "Now the unthinkable has happened to them. And this is not a Third World developing country. This is Europe" (Watson, 2022).

Daniel Hannan was even more blunt in his *Telegraph* article: "They look so much like us. That is what makes it so shocking. Ukraine is a European country. Its people watch *Netflix* and have Instagram accounts, vote in free elections, and read uncensored newspapers. War is no longer something that happens to impoverished and remote populations" (Bayoumi, 2022).

4.3   *The More You Read, the Less You Feel*

Most people in the West believe that the term propaganda only refers to the media and information practices of authoritarian regimes such as those in Russia, North Korea, and Iran. In contrast, the media in Western nations are expected to be trustworthy and reliable. The press is deemed as the "backbone of democracy" because it provides the valuable service of educating citizens and defending the plurality of opinions and freedom of expression. The media in Western democracies are also viewed as "guardians of power" functioning as objective and impartial watchdogs committed to exposing lies, corruption, and injustices.

This naïve view blinds Western individuals to a far less edifying reality. Although Western media is not as tightly controlled as it is in authoritarian regimes, it is still far from free. The manipulation of public opinion is more sophisticated and insidious than the crude propaganda of authoritarian states, but it can arguably be more effective in misleading the public. Rather than relying on outright disinformation, Western propaganda often marginalizes or omits certain issues and buries inconvenient facts until they virtually disappear. This subtle yet pervasive and persuasive type of manipulation serves as a shield against deeper scrutiny.

In *Manufacturing Consent: The Political Economy of Mass Media*, Edward Herman and Noam Chomsky investigated media strategies designed by powerful elites to control what the population sees, hears, and thinks. According to this perspective, the goal of propaganda is to set the boundaries of public discourse, ensuring that the masses debate only those issues and viewpoints deemed "acceptable" by those in power (Herman & Chomsky, 2010).

This control is enforced through a series of "news filters:" media ownership, funding sources, dependence on advertising, and reliance on government or corporate experts. These filters determine which facts are highlighted, which are downplayed, and which are erased altogether. The filters create the illusion of open debate while narrowing the spectrum of discussion so that even opposing views share the same flawed assumptions. One of the most controversial filters adopted by mainstream Western media, which Herman and Chomsky call "worthy and unworthy victims", explains why some lives and tragedies receive ample coverage while others never make the news (Herman & Chomsky, 2010).

There is evidence that Western citizens are misinformed about significant geopolitical facts and historical events. For example, a 2003 study found that most Americans held at least one false belief about the Iraq War such as believing that Iraq was linked to al-Qaeda, that weapons of mass destruction were found, or that the world supported the U.S. invasion. Of course, none of this was true. However, belief in these falsehoods was widespread and correlated with the type of media most consumed. An astonishing 80% of *Fox News* viewers believed at least one falsehood, compared to 71% of CBS viewers, 61% of ABC viewers, and 55% of both NBC and CNN viewers (Kull et al., 2003).

Disinformation about the historical context and the geopolitical events characterizing Israel's occupation of Gaza is also widespread. This misinformation began long before October 7. A 2002 study by the *University of*

Glasgow found that 91% of young British news viewers mistakenly believed that Palestinians were the illegal settlers, rather than the occupied population (Philo, 2002).

These distortions of information affect more than just our thoughts. They also alter how we empathize. By highlighting the suffering of "worthy" victims and erasing that of others, the media modulates our compassion. It decides for us who deserves help and who does not, who deserves to live and who deserves to die.

5   When Will the Night End?

On October 15, 2024, a video capturing the horror of the Gaza genocide went viral. It showed a young man engulfed in flames after an Israeli airstrike near the al-Aqsa Martyrs' Hospital compound in Deir al-Balah. It was Sha'ban al-Dalou, a student at al-Azhar University, killed in the attack along with his mother.

Sha'ban burned alive during the attack. Like many other civilians, after evacuating multiple areas of Gaza, he had taken refuge in a tent near the hospital before Israel turned it into an inferno.

Sha'ban died in full view of the world, which would have remained largely unaware of his tragedy if a bystander nearby had not managed to film his final terrifying seconds of life. No one will ever know what Sha'ban's last thoughts were. He had survived another Israeli airstrike on a mosque that killed twenty people the week before. But this time he could not escape the Israeli killing machine.

The video of Sha'ban al-Dalou's death is hard to forget. It showed people running in terror and screaming their horror. There, engulfed in flames, was a body writhing and crackling. His arm was still raised, as if reaching for help, and an intravenous line was dangling (Witus, 2024).

Another story has broken through the wall of Western media indifference to Palestinian suffering. It concerns the final hours of Hind Rajab. On January 29, 2024, Hind's family made the difficult decision to flee Gaza City's Tel al-Hawa neighborhood in a desperate attempt to escape the ongoing conflict. Hind, aged five, was one of seven family members who piled into the car, including her mother's uncle, Bashar Hamada, his wife, Ana'am, and their four children: Layan, Raghad, Sarah, and Mohammad.

On the same day, an Israeli tank hit the black Kia Picanto in which Hind was traveling with her family. Trapped in the car with her deceased family members before she was killed, she made a deeply moving and tragic phone call.

She pleaded: "Please come get me … I'm so scared. Please call someone to come and take me." Hind stayed on the phone with some *Palestinian Red Crescent* operators for three long hours.

It is only thanks to that call that we know her last words and what happened. Imagine her last seconds, alone and wounded in the car, surrounded by the lifeless bodies of her family members.

Two paramedics in a clearly visible ambulance were given safe passage by the Israeli army and tried to reach her. However, the tank opened fire on the ambulance too, reducing it to a pile of metal sheets. Like Hind, the two rescuers were murdered.

Twelve days later, when the Israeli military temporarily left the area, Hind's small body was found still in the car.

In a documentary about the massacre entitled *The Night Won't End*, Hind's younger brother, Eiyad, who had refused to get into the car and had been allowed to stay with the other family members, asked his mother: "When will the night end?"

"The night will not end," she replied (Al Jazeera, 2024b). There is a really moving moment in Hind's phone call. "Hanoud, why aren't you speaking?" asked the Red Crescent operator.

"I am not speaking because my mouth is bleeding" she replied. "Wipe it with your hand and then tell me if you're still bleeding."

Hind replied: "I don't want to get my shirt dirty, so I don't trouble my mom."

Unsurprisingly, the Israeli army denied any wrongdoing. They justified the bombing that left Sha'ban burned alive as an operation against Hamas. According to Israeli officials, Hamas uses hospitals and refugee camps to hide its terrorist intentions and activities. As for Hind's death, which *ABC News* carefully described in the passive voice, without indicating who killed her in an article entitled "Moments Leading Up to Death of 5-Year-Old Hind Rajab," (ABC News, 2024) the Israeli army claimed that its forces were not present in the area at the time of her death.

The British research organization *Forensic Architecture* exposed this falsehood. They gathered solid evidence proving her killing. According to their investigation, which used satellite imagery and visual evidence, Hind's car was shot 335 times. The Israeli soldiers were so close that they could not have failed to see her inside the vehicle (Forensic Architecture, 2024a).

Another *Sky News* investigation, also based on satellite imagery from January 29, 2024, the day of the attack, confirmed the presence of at least 15 military vehicles in the Tel al-Hawa neighborhood, where the car with Hind was located. According to the author of the investigation, the IDF's presence in the area was undeniable. He then added: "What has happened and what secrets lie beneath the rubble may never be known" (Stuart, 2024).

We may never know the secrets lying beneath the rubble. What we do know, however, is that Israel and its army have been lying without shame throughout the genocide.

In September 2025, I accompanied my mother to the premiere of the film "The Voice of Hind Rajab" in Venice. It was not just a film, but an act of remembrance, a lucid and moving cry about Hind's murder. At the end of the screening, the theater was transformed into pure emotion: 24 minutes of uninterrupted applause, Palestinian flags waving, and chants of "Free, free Palestine." At the time, my mother and I didn't realize it, but we were witnessing the longest standing ovation ever recorded in the festival's history.

None who heard Hind plead for help will ever forget her voice. Her words will remain with us, leaving a void like a black hole in our hearts.

CHAPTER 6

The Herd of Free Thinkers

The most potent weapon of the oppressor is the mind of the oppressed.

STEVE BIKO

∴

"Genocide", from the Greek *genos* (race or tribe), and the Latin root of -cide (to kill), is a deeply contested term. It carries not only profound moral weight but also legal implications. Since South Africa filed a case against Israel at the *International Court of Justice* (ICJ) in The Hague for alleged violations of the Genocide Convention in Gaza, the legal definition has taken on even greater importance.

On January 26, 2024, the *ICJ* issued an order on provisional measures in the case. In its report, the Court stated: "In the Court's view, the facts and circumstances mentioned above are sufficient to conclude that at least some of the rights claimed by South Africa and for which it is seeking protection are plausible. This is the case with respect to the right of the Palestinians in Gaza to be protected from 'acts of genocide'" (ICJ, 2024, p. 23).

Since then, multiple states have formally supported or sought to intervene in the case. These include, among others, Spain, Belgium, Ireland, Mexico, Bolivia, Nicaragua, Colombia, Chile, Cuba, Egypt, Libya, Turkey, and the Maldives (Al Jazeera, 2024a).

Despite the Court's initial findings, the *ICJ* is still in the process of gathering evidence. A final verdict may take years to reach. However, this raises a critical question: does the absence of a final ruling prevent the use of the term genocide to describe what has been happening in Gaza?

It does not.

There are compelling moral and empirical reasons to adopt the term even before the Court delivers its final judgment. While stopping short of a final verdict, the ICJ acknowledged the plausibility of the South African case and ordered Israel to take all measures within its power to prevent acts prohibited by the Genocide Convention. Furthermore, legal formalism cannot obscure the overwhelming reality. The facts on the ground matter far more than the legal technicalities and sophistries. As already made clear, Gaza has experienced an unprecedented death toll, particularly among children, and the fastest drop in life expectancy in recent history. These are not ambiguous matters open to interpretation. They are measurable effects of Israel's onslaught, counted in human lives lost.

There is also incontrovertible evidence of the existence of several mass graves. This consolidates the argument even further. The existence of mass graves has long been a critical factor in international genocide cases. For instance, in the 1995 Srebrenica massacre, the discovery of mass graves played a pivotal role in determining whether genocide had actually occurred.

According to *Reuters*, at least two Gaza hospitals have revealed mass graves containing 400 bodies, some with their hands bound, indicating clear violations of international humanitarian law (Deutsch & van den Berg, 2024). Six more mass graves have since been discovered, including one containing fifteen rescue workers (eight *Red Crescent* workers, six from the *Palestinian Civil Defense* and one *UNRWA* staffer) who were executed "one by one" while attempting to save civilians (Lukiv & Gritten, 2025).

Despite the overwhelming evidence proving the genocide, most Western institutions and some intellectuals have engaged in dialectical acrobatics to deny the obvious. In an editorial titled *Charging Israel with genocide makes a mockery of the ICJ*, the London-based magazine *The Economist* claimed that South Africa's case "cheapens the term" (The Economist, 2024). Just a few years earlier, the same magazine published a piece titled *What is genocide and is Russia committing it in Ukraine?* using a markedly different tone (The Economist, 2022).

The double standard is glaring. Gaza's civilian death toll, both in absolute numbers and in the ratio of civilians to combatants, far exceeded Ukraine's. Yet, this disparity did not appear to trouble *The Economist*. At least not at the time. By dismissing the *ICJ*'s intervention as "mockery," the British magazine did not undermine the court, but its own credibility.

There have also been notable cases of censorship. In early June 2024, the *Columbia Law Review*'s website was taken down by its board of directors after student editors refused the board's request to stop publishing an article by Palestinian scholar Rabea Eghbariah. Titled *Toward Nakba as a Legal Concept*, the manuscript argued that the Palestinian Nakba should be developed as a new legal framework related to, but distinct from, other processes defined in modern international law, including apartheid and genocide. The previous year, the *Harvard Law Review* had refused to publish a similar shorter article it had solicited from Eghbariah, even after the article had been initially accepted and fully edited and fact-checked (Democracy Now!, 2024).

1   **"Inescapably Genocidal"**

In stark contrast to some Western media outlets and intellectuals who have denied or downplayed the scale of Israel's assault on Gaza, genocide scholars have shown a broad consensus. In August 2025, the International Association of Genocide Scholars (IAGS) adopted a resolution declaring that Israel's policies in Gaza met the legal definition of genocide (IAGS, 2025). The following month, the UN Commission of Inquiry reached the same conclusion.

Leading experts had recognized this much earlier. A few months before, Melanie O'Brien, president of IAGS, stated that Israel's assault on Gaza amounted to genocide (Sondos, 2025).

Omar McDoom, an associate professor at the London School of Economics, identified three key markers that support the genocide hypothesis: a) the deliberate, organized, and sustained nature of the violence; b) the targeting of a clearly identifiable civilian (ethnic) group; c) concrete steps taken to prevent the group's survival or reproduction. As McDoom concluded, "Using these markers, it could be argued that the actions of the Israeli government are genocidal" (McDoom, 2024).

John Quigley, professor emeritus at The Ohio State University, and author of *The Genocide Convention: An International Law Analysis*, shared a similar view. In a piece published in March 2024, he focused on one notable aspect of genocide liability: "deliberately inflicting conditions of life calculated to bring about physical destruction in whole or in part." He also compared the case of genocide in Bosnia with that of Gaza: "the factual situations in Gaza differ markedly from those in Bosnia. In the latter, 'conditions of life' were being inflicted on discrete numbers in particular towns or detention centers, whereas in Gaza, they were being inflicted on nearly the entirety of the population." Quigley added: "In Bosnia, the overall population had a physical option to flee the conditions, whereas in Gaza they did not. The Gazans, on the contrary, were being forced into the very conditions that were calculated to bring about their destruction" (Quigley, 2024).

Dirk Moses, an Australian professor at the City University of New York and the editor-in-chief of the *Journal of Genocide Research*, focused his attention on Israel's use of an artificial intelligence system that tracks the locations of Hamas and Palestinian Islamic Jihad members when they are at home with their families and neighbors. This allows Israeli forces to bomb them there. As he observed: "It's not unintentional, and it's not even collateral, it's calculated," he said. "And when you marry that with the public statements of Israeli leaders, then it's impossible to ignore that there's a fusion of military and genocidal logics" (Cohen, 2024).

Of course, some realized it was genocide sooner than others. Just a few days after October 7, 2023, more than 800 scholars and practitioners in international law, conflict studies, and genocide research signed a public statement arguing that the actions of Israeli forces could constitute genocidal acts (TWAILR, 2023).

In late July 2025, Jonathan Montomoli, Ghassan Abu-Sittah, Ilan Pappé, and I published a short article in *The Lancet* entitled *Break the Selective Silence on the Genocide in Gaza* (De Vogli et al., 2025). The paper discussed an open letter that prompted numerous academic associations to officially recognize the genocide. The article received some media coverage in Italy. However, during an interview, a local radio presenter asked me, "Why so late?" He was unaware that, on October 20, 2023, I had written an article for *Il Fatto Quotidiano* titled *The Massacre of Innocents in Gaza: Can We Accuse Israel of Genocide?* (De Vogli, 2023).

The piece was met with silence, outrage and personal attacks. It was too much and too soon. As the situation in Gaza worsened, however, more experts realized the severity of what was unfolding.

Amos Goldberg, professor of Holocaust history at the Hebrew University of Jerusalem, wrote a powerful article with a self-explanatory title, "Yes, it is genocide." He stated: "At the end of six months of brutal war, it is no longer possible to escape this conclusion. Jewish history will henceforth be stained with the mark of Cain, the 'crime of crimes,' which cannot be erased from its forehead. As such, it will be on trial for generations" (Goldberg, 2024). William Schabas, author of *Genocide in International Law: The Crime of Crimes*, made clear, during an interview for *Der Spiegel*, that "there is a very strong case for arguing that Israel's response constitutes the crime of genocide" (Prosinger, 2024).

Lee Mordechai, a historian at the Hebrew University of Jerusalem, wrote: "The enormous amount of evidence I have seen ... has been enough for me to believe that Israel is currently committing genocide against the Palestinian population in Gaza" (Mordechai, 2024).

Maung Zarni, another genocide scholar and human rights activist, and Nobel Peace Prize nominee, put it even more bluntly: "What we are seeing in Gaza is a repeat of Auschwitz ... mass extermination without the gas chambers" (Kartal, 2024).

"Genocide in plain sight": this is how Gregory Stanton, founder of *Genocide Watch*, former research professor of genocide studies and prevention at George Mason University, and past president of the IAGS, described Israel's actions in Gaza (Sarfraz, 2024).

Even scholars who initially resisted labeling the tragedy of Gaza as genocide have since changed their stance. In an editorial in *The Guardian*, former IDF soldier and genocide historian Omer Bartov publicly reversed his earlier position. "On November 10, 2023, I wrote in *The New York Times*: 'As a historian of genocide, I believe there is no evidence that genocide is now taking place in Gaza, although it is very likely that war crimes and even crimes against humanity are taking place. [...].'" Bartov then added: "I no longer believe that. By the time I travelled to Israel, I was convinced that, at least since the IDF attack on Rafah on May 6, it is no longer possible to deny that Israel is systematically committing war crimes, crimes against humanity and genocidal acts" (Bartov, 2024).

In addition to public statements and legal assessments, the academic literature contains numerous contributions that explicitly describe the situation in Gaza as genocidal. The *Journal of Genocide Research*, the oldest and most prominent periodical in the field, has devoted significant coverage to the topic. As Abdelwahab El-Affendi, a professor of politics at the Doha Institute for Graduate Studies, wrote in the same journal, "the genocidal intent and consequences of the Israeli assault are becoming indisputable by the day" (El-Affendi, 2024).

Nimer Sultany, at the *School of Oriental and African Studies* (SOAS) in London, published an article arguing that the violence in Gaza met the definitional threshold of genocide under international law and urged the international community to take action to prevent further atrocities (Sultany, 2024).

In *We are Fighting Nazis: Genocidal Fashionings of Gaza(ns) After 7 October*, Zoe Samudzi, visiting assistant professor at the Strassler Center for Holocaust and Genocide Studies at Clark University, concluded that Israel has committed "nearly every act outlined in Article II [of the Genocide Convention] ... that accounts for the more totalized 'destruction of the national pattern of the oppressed group'" (Samudzi, 2024).

Of those who sounded the alarm first, Martin Shaw, a professor of international relations at the Barcelona Institute of International Studies and author of *What Is Genocide?* and *The New Age of Genocide*, deserves special recognition. He published *Israel, Gaza and the Spectre of Genocide* on October 13, 2023. Within months, he followed with a piece that left little doubt about what was unfolding in Gaza: *Inescapably Genocidal* (Shaw, 2024).

1.1   *As If There Were No Humans*

The consensus around the genocide in Gaza is not limited to genocide scholars. Leading human rights organizations, UN Special Rapporteurs and non-governmental organizations (NGOs) have also sounded the alarm, describing Israel's actions in unequivocal terms.

The *Lemkin Institute for the Prevention of Genocide*, named after Raphael Lemkin, a well-known Polish lawyer of Jewish descent who coined the term "genocide", issued one of the earliest and clearest statements. After 18 months of frustration due to their repeatedly ignored warnings, the organization stated: "We have warned of genocide since October 13, 2023. It was clear then and is clear now. Israel is committing genocide. The US is complicit" (Lemkin Institute, 2025).

Similarly, a detailed legal analysis by the *University Network for Human Rights* concluded unequivocally: "Israel has committed acts of genocide, namely killing, causing serious harm, and inflicting conditions of life calculated and intended to bring about the physical destruction of Palestinians in Gaza."

Defence for Children International, a reputable NGO with a long history of child advocacy in war zones, also issued a decisive statement concluding that Israeli forces have been carrying out a "genocide against Palestinian children" (Defense for Children Palestine, 2023).

The term "genocide" has also been deliberately and explicitly used in official reports by various United Nations experts. Francesca Albanese, the *UN Special Rapporteur on the Situation of Human Rights in the Palestinian Territories*, titled her report *Anatomy of a Genocide*. In the document she concludes: "By analyzing the patterns of violence and Israel's policies in its assault on Gaza, this report finds reasonable grounds to believe that the threshold indicating the commission of genocide by Israel has been met" (UN Human Rights Council, 2024). Michael Fakhri, the *UN Special Rapporteur on the Right to Food*, was also unequivocal: "Israel's campaign of starvation against the Palestinian people in Gaza constitutes genocide." Fakhri, an expert in international economic and food law, stressed the use of starvation as a deliberate weapon of war (Middle East Monitor, 2024a). Tlaleng Mofokeng, the *UN Special Rapporteur on the Right to Health*, accused Israel of "knowingly and intentionally imposing famine, prolonged malnutrition and dehydration" (United Nations, 2024b). In a January 2025 press release, together with Francesca Albanese, she stated: "For well over a year into the genocide, Israel's blatant assault on the right to health in Gaza and the rest of the occupied Palestinian territory is plumbing new depths of impunity" (OHCHR, 2025a).

In December 2024, *Amnesty International* and *Human Rights Watch* released two official reports stating that Israel committed "genocide" in Gaza. *Amnesty International* published a devastating document titled *You Feel Like You Are*

Subhuman: Israel's Genocide Against Palestinians in Gaza. The report accuses Israel of committing at least three of the five acts prohibited by the 1948 Genocide Convention: the indiscriminate killing of civilians, causing serious bodily or mental harm, and imposing living conditions aimed at destroying the Palestinian population. The report also details the unprecedented level of destruction Israel has perpetrated with total impunity (Amnesty International, 2024).

Human Rights Watch, too, found Israel guilty of genocide. In a report, titled *Extermination and Acts of Genocide: Israel Deliberately Depriving Palestinians in Gaza of Water*, the U.S.-based organization stated that Israeli authorities have deliberately deprived Palestinians of access to safe drinking water and sanitation. The report states: "This isn't just negligence; it is a calculated policy of deprivation that has led to the deaths of thousands from dehydration and disease. It is nothing short of the crime against humanity of extermination, and an act of genocide" (Human Rights Watch, 2024c).

A few days after *Amnesty International* and *Human Rights Watch* published their reports, *Doctors Without Borders* released a document highlighting how repeated Israeli military attacks on Palestinian civilians and the systematic denial of humanitarian aid have literally destroyed the basic conditions of life in Gaza. According to the report, there were "clear signs of ethnic cleansing as Palestinians are forcibly displaced, trapped, and bombed." The document also emphasized the destruction of the healthcare system and the unprecedented number of deaths and detentions of medical personnel (Doctors Without Borders, 2024c).

2   Believing the Unbelievable

Imagine a group of hooligans staging a racist rampage in a major European city, vandalizing property and chanting songs with lyrics like: "There are no schools in Gaza because there are no children." Their actions provoke widespread outrage, even retaliatory violence. While such reactions are never justifiable, the cruelty and explicit provocation have played a significant role in triggering them.

Then, the prime minister of the country from which these hooligans hail, himself indicted by the ICC for war crimes, seizes the moment. With remarkable political agility, he reframed the incident as a "pogrom" (massacre) and even compared it to Kristallnacht, likening the backlash against a bunch of violent, racist agitators to a Nazi purge. Under the pressure of government-aligned propaganda, much of the mainstream media fell into line, recasting this mob of supremacist hooligans as victims. And anyone who dared to criticize or confront them with facts is swiftly labeled as an anti-Semite.

Let us recap: you publicly and gleefully celebrate the extermination of over 18,000 children, singing about it, laughing about it. Yet, those who protest your cruelty, who call out your glorification of mass killing, are the ones branded as racists. Those who reject your narrative, because they refuse to be gaslighted, or to believe the unbelievable, are cast as villains.

What kind of dystopian, Orwellian world is this, where truth and lies are so thoroughly inverted? What Kafkaesque society makes you feel like a liar, a fool, or even a criminal for simply speaking a plain and evident truth?

It may sound like fiction. But it is not. This happened in the so-called "free world", a society that prides itself on democracy, freedom of expression and defense of human rights. Fortunately, not every voice in the Western press has surrendered to the machinery of power. While some media outlets, such as Italy's *La Repubblica*, echoed the Israeli Prime Minister's baseless claims by calling the confrontation a "pogrom" (Tercatin, 2024), others, like Finland's *Helsinki Times*, chose factual reporting. They provided context, described the provocations that led to the incident and did not invert the victim and the perpetrator of hate crimes (Helsinki Times, 2024).

No single media outlet defines a nation's relationship with truth, but the contrast is revealing. According to the 2024 *World Press Freedom Index*, which ranks countries based on the degree of liberty journalists enjoy from censorship, political interference, and legal harassment, Finland ranks 5th in the world. Italy ranks 46th (Reporters Without Borders, 2024).

A coincidence? Maybe, but the index closely correlates with the quality of information and public trust in media across the 180 countries it assesses. Now, some might raise the familiar, and overused, refrain: correlation is not causation. Fair enough. But causation does not even show up without it.

2.1   *Algorithmic Gaslighting*

In March 2025, internal documents obtained by *Drop Site News*, an independent investigative journalism outlet, exposed what may be the largest mass censorship operation in modern history. The revelations centered on *Meta*, the multinational tech giant that owns *Facebook*, *Instagram*, and *WhatsApp*. *Meta* conducted an extensive campaign of content suppression, executed at the behest of the Israeli government. The campaign consisted of targeting social media posts critical of Israel or even mildly supportive of Palestine (Ahmed, 2025).

According to the leaked data, *Meta* complied with 94% of Israel's takedown requests since October 7, 2023, making Israel, by far, the leading source of online censorship. Automated censorship surged, resulting in 38.8 million content moderation interventions, including takedowns and algorithmic suppression. In one striking example, over 90,000 posts were removed within thirty seconds. Alarmingly, many of these takedowns involved vital health, humanitarian, and human rights information, content that could have helped protect lives, documented war crimes, or aided displaced Palestinians.

This is not an isolated incident. In late 2023, a *Human Rights Watch* report documented over 1,050 instances of censorship by *Meta* targeting pro-Palestinian content. This included peaceful expressions of solidarity, documentation of war crimes, and updates on the conditions of life in Gaza and the West Bank. Of those cases, 1,049 involved pro-Palestinian content. Only one regarded content supportive of Israel. While Meta does allow some limited criticism of Israeli policies, *Human Rights Watch* concluded that its systemic suppression of Palestinian voices represents a clear violation of freedom of expression and the right to access information, universal rights protected by international law (Human Rights Watch, 2023).

But censorship is only one side of this strategy. The other is disinformation. Since October 7, hundreds of inauthentic accounts have flooded platforms like *X*, *Facebook*, *Instagram*, and *Reddit* to amplify pro-Israel narratives, especially aimed at North American audiences. These coordinated campaigns routinely downplayed or denied Israeli human rights violations, spread Islamophobic rhetoric, framed university campuses as "unsafe" for Jewish students, and equated pro-Palestinian protests with anti-Semitism (Sadek & Digital Forensic Research Lab, 2024).

Some of these accounts also fueled a targeted disinformation campaign against the United Nations Relief and Works Agency (UNRWA), falsely accusing its staff of involvement in the October 7 attacks. These claims were amplified by prominent articles in *The New York Times* and *The Wall Street Journal*, both relying heavily on an Israeli intelligence dossier (Bergman & Kingsley, 2024; Luhnow, 2024).

However, subsequent investigations by *Channel 4 News* and *The Financial Times* cast serious doubt on the credibility of the dossier, citing unsubstantiated claims and a lack of verifiable evidence. A later report from UNRWA revealed that several of its staff had been detained and tortured by Israeli authorities, coerced into false confessions implicating the agency in the October 7 attacks (Reuters, 2024a).

This concerted campaign of disinformation and censorship appears to be closely tied to Israel's broader strategy to influence public opinion in countries that fund its military operations, especially the United States, Canada, and EU member states. An investigation by the Israeli watchdog group *FakeReporter* uncovered five propaganda websites linked to Stoic, a political consulting firm contracted by Israel's Ministry of Diaspora Affairs. The firm was reportedly paid $2 million to influence at least 128 U.S. Congressmen and state legislators to support Israel's actions in Gaza (Frenkel, 2024).

2.2   *"Anti-Semitic" Evidence*

Anti-Semitism is a specific form of racism defined as hatred or prejudice against Jews and Judaism. It has left scars on history, culminating in the Holocaust, one of the most horrific genocides ever perpetrated, if not the most horrific. Every year, we appropriately commemorate this tragedy through education, media campaigns, and initiatives like *Holocaust Remembrance Day.*

Anti-Semitism, like Islamophobia and other forms of racism, remains a persistent plague, and its resurgence, especially after October 7, is deeply troubling. This form of hatred has deep historical roots, particularly in Western societies. The increase in anti-Semitic attacks since October 7, exacerbated the collective trauma of a people who have endured centuries of persecution and the lasting effects of the Holocaust. True anti-Semitism is a despicable phenomenon that must be accurately identified and unreservedly condemned.

However, it is not only anti-Semitism to be identified and condemned, but also its cynical manipulation. In a rational and intellectually honest debate, the difference between real anti-Semitism and legitimate outrage over Israel's human rights violations would be crystal clear. Yet, many organizations tasked with combating anti-Semitism have adopted a politically convenient interpretation of the term. They use it to stigmatize any dissent against Israel's policies, no matter how reasonable, evidence-based, or compassionate.

The weaponization of anti-Semitism, stemming from a distorted reality based on biased nationalist beliefs, does not protect Jews from hatred; rather, it protects Israel from accountability. It functions as a powerful tool to silence writers, academics, journalists, and even Jewish critics (Landy et al., 2020). The weaponization of anti-Semitism has also been used to launch organized attacks against progressive political leaders that showed solidarity toward Palestine. British politician Jeremy Corbyn is a notable example (Winstanley, 2023).

During an interview on *Democracy Now!* on August 14, 2002, Shulamit Aloni, a former Israeli education minister and head of the Meretz party, candidly acknowledged: "It's a trick. We use it all the time." She continued: "When someone from Europe criticizes Israel, we bring up the Holocaust. In the United States, when people criticize Israel, they are anti-Semitic" (Democracy Now!, 2002).

Since October 7, a very long list of individuals, organizations and even countries have been stigmatized as anti-Semitic. Those accused include Pope Francis, U.N. Secretary-General António Guterres, and Pink Floyd musician Roger Waters. International bodies such as the *ICC, ICJ, UNICEF, WHO, UNESCO, Amnesty International, Human Rights Watch* and *Doctors Without Borders*, have also been labeled as such. Over 150 countries that have either condemned Israel's blatant violations of international law or recognized Palestinian statehood must also be on the list.

At times, even data appear to be branded as anti-Semitic.

What about a research analysis published in the *British Medical Journal* documenting Israel's unprecedented killing of children in war zones (Bhutta et al., 2024)? Anti-Semitic?

And a study in *The Lancet* revealing an unprecedented catastrophic decline in life expectancy in Gaza (Guillot et al., 2025)? Clearly, *The Lancet* must be anti-Semitic too, right?

What about a quantitative analysis published in *Frontiers in Public Health* demonstrating that civilians are the "... object of war ... the primary focus of the conflict" (Ayoub et al., 2024)? Also anti-Semitic?

Even parameters widely used by military analysts, human rights observers, and international courts to assess compliance with the laws of war, such as the ratio of civilian to combatant casualties, must not be immune from accusations of anti-Semitism. According to the *Uppsala Conflict Data Program*, approximately 10% of fatalities in Ukraine since the Russian invasion began in February 2022 were civilians, while 90% were combatants. In Gaza, however, the proportions are nearly reversed, with 83% of those killed being civilians (Graham-Harrison et al., 2025). These numbers explain the difference between war and genocide.

Russia has faced more than 16,000 sanctions. How many sanctions has Israel faced in almost two years of genocide? Virtually zero!

3   Jewish Voices for Truth and Justice

Perhaps the most preposterous tactic in the weaponization of anti-Semitism is the branding of Jewish thinkers, scholars, and Jewish human rights organizations that criticize the occupation, apartheid, and genocide in Gaza as "self-hating Jews," while labeling people who engage with their work, and draw intellectual inspiration from them, as anti-Semites.

Many of today's most insightful and compassionate critics of Israeli policies are Jewish. Guided by ethical convictions and a commitment to truth and justice, these individuals have become global moral leaders, truth-tellers who challenge propaganda and reveal the stark realities unfolding in Gaza. Yet rather than being honored for their clarity and conscience, they are often vilified, smeared and attacked.

The slur "self-hating Jew" has been directed at some of the most respected intellectuals of our time. Noam Chomsky, a professor emeritus at MIT and one of the most cited scholars in history, has spent decades dismantling state propaganda and defending Palestinian rights. His works, including *Gaza in Crisis* and *On Palestine*, are cornerstones of critical thought on foreign policy and Israeli militarism.

Avi Shlaim, an Israeli historian and Oxford professor, has made profound contributions to our understanding of Israel's political history. In books such as *The Politics of Partition* (Shlaim, 1998) and *War and Peace in the Middle East*, Shlaim makes clear that anti-Semitism is a historically European phenomenon, not an Arab one. This shatters a myth that is commonly used to justify Israeli violence (Shlaim, 1995).

Historian and University of Essex professor Ilan Pappé too has challenged dominant narratives with his meticulous scholarship. In *The Ethnic Cleansing of Palestine* (Pappé, 2007) and *Ten Myths About Israel*, he exposed the historical distortions used to legitimize occupation and erase Palestinian suffering from Western memory (Pappé, 2017).

These are not just a few outliers. Countless Jewish writers, journalists, and organizations reject Israeli state violence and show solidarity with the oppressed.

Few scholars embody moral clarity and intellectual courage like political scientist Norman Finkelstein. The son of Holocaust survivors, he has been blacklisted from academia because of his determination to speak the truth when others dared not.

In his seminal book, *The Holocaust Industry*, Finkelstein argues that the memory of Jewish suffering has been cynically weaponized to deflect criticism and justify the oppression of Palestinians (Finkelstein, 2003). In a famous speech delivered at the *University of Waterloo* Finkelstein offered a brutal yet truthful reflection after being challenged by a young female protester in the audience. "Every single member of my family on both sides was exterminated.

Both of my parents participated in the Warsaw Ghetto Uprising. It is because of the lessons my parents taught me and my siblings that I cannot remain silent when Israel commits its crimes against the Palestinians." As the protesters grew louder, Finkelstein added: "There is nothing more despicable than using their suffering and martyrdom to justify the torture, brutalization and home demolitions that Israel inflicts on Palestinians every day" (Ridgen & Rossier, 2009).

Noam Chomsky, Avi Shlaim, Ilan Pappé, and Norman Finkelstein are among the many contemporary Jewish intellectuals committed to preventing the victims of one genocide from becoming the perpetrators of another.

Two of the most iconic figures in this tradition are Albert Einstein and Hannah Arendt. On December 2, 1948, they and 26 other Jewish signatories published a letter in *The New York Times* condemning the rise of fascist ideology in the early days of Israeli politics. The letter stated: "Among the most disturbing political phenomena of our time is the emergence in the newly created State of Israel of the "Freedom Party" (*Tnuat Haherut*), a political party closely related in its organization, methods, political philosophy, and social appeal to the Nazi and Fascist parties. It was formed from the members and followers of the former Irgun Zvai Leumi, a terrorist, right-wing, chauvinist organization in Palestine" (Einstein & Arendt, 1948).

Though forgotten by many and suppressed by some, this letter is yet another reminder that some of the most powerful indictments of Israeli state violence have come from within the Jewish community itself. These Jewish thinkers have done more than critique; they have deprogrammed public opinion. They have cleared the fog of propaganda, challenged moral hypocrisy, and inspired a vision of justice rooted in critical thinking and shared humanity.

Ironically and tragically, reading and quoting these Jewish scholars is now grounds for being labeled an anti-Semite. But their words remain clear and unflinching, reminding us that standing with the oppressed is a moral imperative regardless of who the oppressor is.

4　Stenographers of Power

The genocide in Gaza has become more than a humanitarian catastrophe. It has become a global battlefield for truth, testing the integrity, compassion, and intellectual honesty of the Western intelligentsia. With brutal clarity, it has exposed the moral divide between courage and cowardice, between ethical responsibility and self-serving careerism.

As Edward Said noted in the *Reith Lectures*: "Nothing, in my view, is more reprehensible than those habits of mind in the intellectual which produce avoidance, that characteristic turning away from a difficult and principled position which one knows to be right, but which one chooses not to take." Said added: "You do not want to appear too political; you are afraid of appearing controversial; you want to maintain a reputation for being balanced, objective, moderate ... If anything can denature, neutralize, and ultimately kill a passionate intellectual life, it is the internalization of such habits." According to Said, despite the vilification endured by those who advocate for Palestinian rights, the role of the intellectual remains clear, "the truth deserves to be spoken, represented by an unafraid and compassionate intellectual" (BBC, 1993, p. 4).

Yet, in the face of mass death, displacement, and the destruction of Gaza, most Western intellectuals and journalists looked away. They remained silent, not necessarily out of ignorance but out of fear, fear of controversy, fear of professional consequences, fear of being labeled, marginalized, vilified, or attacked.

Some practiced self-censorship, carefully avoiding "unbalanced" positions. Others didn't need to. Their worldview is so deeply shaped by colonial assumptions, nationalist mythologies, and strategic alliances that Gaza barely registers as a moral concern. Many defended Israel's actions by repeating myths that sought to sanitize the language of war, justify the unjustifiable, and rationalize the indefensible.

In an iconic exchange, former BBC journalist Andrew Marr asked Noam Chomsky, "How do you know I'm self-censoring? How do you know that journalists self-censor?" Chomsky replied, "I'm not saying you're self-censoring. I'm sure you believe everything you say. But what I'm saying is that if you believed anything else, you wouldn't be sitting where you are" (Media Lens, 2021).

Chomsky's point was not about conspiracy; it was about structural conformity. Intellectuals and journalists in mainstream Western media do not need to be forced to obey. Rather, they internalize the boundaries of what is acceptable to think, say, or publish. Critical thought is permitted, but only within carefully policed, sanitized limits. As long as one stays within those limits, dissent is allowed and even celebrated. Step outside them, however, and the cost becomes clear.

Within this framework, journalists and intellectuals are not guardians of democracy, but servants of privilege. They turn into stenographers of power, maintaining the illusion of a free and critical press while keeping subversive truths outside the frame.

Chomsky famously borrowed a phrase from writer and cultural critic Harold Rosenberg to describe this class of thinkers: "A herd of independent minds" (Chomsky, 2006).

4.1 *Selective Silence*

"Do you know what unsettles me most about this war? Not just the bombs. Not just the starvation. Not just the cold, calculated siege of two million souls. It's the silence," wrote Dr. Ezzideen Shebab, a young medical doctor operating in the ruins of the Indonesian Hospital in Gaza. He then continued: "It's the choking, complicit silence of a world that once branded the 20th century as the 'century of freedom.' ... Children are buried beneath concrete, not by accident, but by design. And the world watches on, not with shock, but with schedule. Press conferences. Ceasefire negotiations. Think pieces. Silence" (Shehab, 2025).

Dr. Shebab has a point. Perhaps, the most disheartening aspect of the Gaza genocide is not the endorsement and justification of Israel's war crimes. It is not the racial hatred and genocidal statements by high-profile figures in Israel. It is not the collective insanity of a hundred Israeli doctors who called for Israel to target Gaza hospitals in an open letter (Middle East Eye, 2023b). No. The most disheartening moral outcome of this genocide has been the deafening silence of so many Western intellectuals.

Among those most strikingly absent from moral leadership were associations of medical and public health professionals whose mission is to protect health and social justice.

Some organizations acted right from the start and did everything they could to stop the genocide. Others remained conspicuously silent (Pagadala & Nichols, 2025). While hospitals were levelled, doctors targeted, and ambulances bombed, most professional bodies remained mute, retreating behind a shield of "neutrality." Their silence was made even more conspicuous by their past eloquence: the same organizations had been unambiguous in their response to the Russian invasion of Ukraine. We called this "selective silence" (De Vogli et al., 2025).

A stark example is the *American Medical Association* (AMA) which expressed "outrage" over "the senseless injury and death inflicted on the Ukrainian people by the Russian army" (Robeznieks, 2022). However, when presented with a resolution to protect civilians and medical personnel in Gaza, the organization declined to support it. While the AMA President was quick to condemn the October 7 attacks on Israel, he remained silent during the relentless massacre of women and children in Gaza (Marques, 2024). The American Public Health Association (APHA) took a similar position.

A group of Australian doctors spoke out against the silence of the Australian Medical Association (AMA) in a March 2024 letter published in *The Lancet*. They expressed disbelief at the indifference of their professional leadership: "When Russia invaded Ukraine in 2022, the *Australian Medical Association* (AMA) swiftly condemned the aggression ... Since October 7, however, Israel has conducted over 600 attacks on health facilities and killed at least 300 health workers ... In just the first week, more children were killed in Gaza than in an entire year of war in Ukraine." The authors then complained: "The Hippocratic principle of first, do no harm, precludes our silence" (Abbas & Mitchell, 2024). In April 2024, three months after the ICJ implied that it was "plausible" to conclude that Israel was committing "acts of genocide," the AMA stated: "The AMA believes that all parties involved in the conflict in Israel and Gaza must respect the principles of medical neutrality during times of armed conflict" (Australian Medical Association, 2024).

How will history remember this silence? How will future generations judge those who chose to remain neutral as children in Gaza were bombed, starved, and erased? Will this silence be remembered as one of the most shameful episodes of intellectual cowardice in history?

Some voices, however, refused to be complicit. They broke the silence not because they were "pro-Palestinian," as some pundits claimed, but simply because they wanted to remain human.

5   When Your Enemy Is the Truth

"If you're reading this, it means I have been killed, most likely targeted, by the Israeli occupation forces. When this all began, I was only 21 years old, a college student with dreams like anyone else. For the past 18 months, I have dedicated every moment of my life to my people. I documented the horrors in northern Gaza minute by minute, determined to show the world the truth they tried to bury" (Conley, 2025a).

These were the final words of Hossam Shabat, a young *Al Jazeera* journalist, before he was killed by Israeli forces in northern Gaza.

Weeks later, it was the turn of journalist Ahmed Mansour. He was killed together with fellow journalist Hilmi al-Faqawi and a civilian, Yousef al-Khazindar, when another Israeli strike targeted a tent sheltering journalists in Khan Younis. Mansour's death received public attention because he was captured on video burning alive. He died after two days of agony.

Several other journalists were wounded in the same attack, some critically. Abed Shaat, a journalist who survived the attack, said that the tent, widely known as a designated shelter for media personnel, was hit at around 3 a.m., without warning. "The tent was known to everyone as one for journalists. This confirms that it was a targeted attack," Shaat emphasized.

The Watson Institute for International and Public Affairs at Brown University released a report titled *News Graveyards: How Dangers to War Reporters Endanger the World* that showed that the Israeli assault on Gaza since October 2023 has killed more journalists than the U.S. Civil War, World Wars I and II, the Korean War, the Vietnam War, the Yugoslav Wars, and the post-9/11 war in Afghanistan combined (Turse, 2025). Similarly, in February 2025, the *Committee to Protect Journalists* reported that the Israeli assault had taken "an unprecedented toll on Gaza journalists," making it "the deadliest period for journalists since data collection began in 1992" (Committee to Protect Journalists, 2025).

To grasp the scale, consider that in the first three years of Russia's invasion of Ukraine, 21 journalists were killed. In Gaza, that number reached 233 by October 2025.

Despite this, Western mainstream media outlets have largely mirrored Israeli government narratives, often resorting to intellectual contortions to justify the attacks. Of course, these justifications collapse under scrutiny. The United Nations Human Rights Office of the High Commissioner wrote in a report: "Despite being clearly identifiable in jackets and helmets marked 'press' or travelling in well-marked press vehicles, journalists have come under attack, which would seem to indicate that the killings, injuries, and detentions are a deliberate strategy by Israeli forces to obstruct the media and silence critical reporting." UN experts have also expressed concern about "restrictions on other journalists from accessing Gaza, combined with severe disruptions of the Internet …[as] major impediments to the right of information of the people of Gaza as well as the outside world" (OHCHR, 2024a).

Western journalists, in many cases, have not only remained silent, but have also played a role in shaping and sanitizing the narrative. As former *New York Times* journalist Chris Hedges observed, although around 4,000 foreign journalists were accredited in Israel to cover the war, virtually none of them conducted on-the-ground investigations. Instead, he recounts how many were taken to staged street performances by the Israeli army and given daily briefings by senior Israeli officials. "Those not accredited", he wrote, "are enemies, terrorists" (Hedges, 2024a).

Indeed, only a handful of journalists stood among the ruins of Gaza. They documented the unspeakable, the unthinkable. Some even paid the ultimate price. Their deliberate targeting proves that they were not collateral damage, but rather stories that never needed to be told. Stories that needed to be silenced.

In an article published in 2002 entitled *Kill the Messenger: Why Palestine Radio and TV Studios are Fair Targets in the Palestine/Israeli War*, American journalist, historian and Pulitzer Prize-winning author, Anne Applebaum wrote, in reference to the destruction of the *Voice of Palestine* broadcasting center: "Although, again, I can't quite see the point of destroying the building—the radio went off the air for a few hours but began broadcasting again from another set of studios—the official Palestinian media is the right place for Israel to focus its ire" (Applebaum, 2002).

There are no words to describe these words.

But as a slogan used by opponents of the Gaza genocide puts it: "You only target journalists when your enemy is the truth."

The World's Conscience

Many of us ask ourselves,
What would I do if I was alive during slavery?
Or the Jim Crow South? Or apartheid?
What would I do if my country was committing genocide?
The answer is you are doing it.
Right now.

∴

These were some of the last words of Aaron Bushnell before he set himself on fire outside the Israeli Embassy in Washington, D.C., on Sunday afternoon, February 25, 2024. Dressed in his U.S. Air Force uniform, he calmly approached the building and ignited himself in protest. As the flames engulfed him, he shouted "Free Palestine" before collapsing to the ground.

He died hours later at a nearby hospital.

It is unfortunate that Western mainstream media did not interpret his act as a tragic yet heroic instance of self-immolation, like that of Jan Palach or the Tibetan monks who set themselves on fire to protest oppression.

Instead, the mainstream media framed his act as possibly indicative of mental illness. A *BBC News* report claimed that his friends struggled to comprehend his death. However, when interviewed, they described Aaron as a "normal, quiet, friendly, quirky guy" who drank root beer, owned a cat named *Sugar*, and loved *The Lord of the Rings*. They described him as an "incredibly strong-willed" person. Aaron was a dedicated soul. He volunteered with homeless charities, embraced left-wing politics and refused to be complicit in Israel's annihilation of Gaza (BBC News, 2024).

Of course, to anyone unable to grasp the deeper meaning of such an act, a protest suicide may seem absurd or even insane. But what if the truly insane are those who did nothing while the Gaza genocide unfolded?

1   Conscientious Protesters

Since the Israeli onslaught began and countless civilians, including women and children, have been slaughtered before our eyes, Gaza has become a mirror of the West's collective moral failure. It has forced a reckoning and revealed the uncomfortable truth about who we are and who we are not (De Vogli & Ferretti, 2024).

Gaza has become the moral blind spot in the West's eye.

Yet it has also turned out to be a litmus test for our humanity. Activists, doctors, scientists, journalists, and ordinary citizens across the planet have organized protests, signed petitions, written and spoken out, creating a chorus of global resistance. They stood firmly against the genocide, relying only on their conscience.

But what does "conscience" really mean? The term is difficult to define. There is no single, unified idea of it. Its definition varies depending on one's perspective and discipline. Historians, philosophers, psychologists, theologians, and many other scholars still struggle to conceptualize it. Some have offered partial demystifications, but the debate remains open and controversial.

Etymologically, the term comes from the Latin *conscientia*, which literally means "shared knowledge." However, it is not just intellectual knowledge. Rather, it refers to a deeply personal moral awareness, often described by metaphors such as an "inner voice" or the "soul." This is the faculty that leads to introspection, self-evaluation, and, ultimately, ethical action.

Although sometimes confused with conscientiousness, a personality trait associated with diligence, orderliness, and dutifulness, conscience goes further. It reflects a sense of right and wrong, often aligned with universal ethical principles such as the Golden Rule. In many spiritual traditions, conscience is considered the wisdom of the soul: a moral compass rooted in humility, sobriety and dignity. Despite their differences, many theologians and philosophers agree that individuals with conscience generally perform just and generous acts, regardless of their reasons.

Conscience does not necessarily arise from political obligation, religious mandate, or social pressure. It often simply comes from within. Conscience moves people to act justly, even when justice seems inconsequential or entails a personal cost. At its core, conscience manifests itself as compassion, an abiding concern for those who suffer. Without being esoteric or too vague, conscience can even resemble a transcendental form of love, a state of heightened awareness in which the individual experiences a profound connection with people, with nature, and the cosmos itself.

Conscience can also be understood as both an inner judge and a guiding sense of duty. Immanuel Kant famously described conscience as "an inner court within man," referring to the unwritten moral laws that guide our most important decisions (Kant & Gregor, 1996).

This inner awareness, however, often gives rise to profound dilemmas. Our internal moral compass, which distinguishes right from wrong, often urges us to act according to principles, even when such actions conflict with the laws of society. This tension becomes particularly acute when those laws are perceived to be unjust.

A question that grips moral philosophy is this: is it right for individuals to always obey authority and established law? Or should they remain faithful to their own sense of justice, especially when they believe that authority is corrupt, or that laws violate fundamental ethical principles?

1.1   *Not in My Name*

Freedom of conscience is often invoked to justify conscientious objection, which is the refusal to perform certain acts required by law or authority for moral or ethical reasons. People often say, "I must follow my conscience," especially in the case of so-called conscientious objectors. The tension between inner moral conviction and obedience to government authority has long fascinated historians, writers, artists, philosophers, political theorists, and religious thinkers.

This debate is often inspired by figures like Henry David Thoreau, who chose jail over paying taxes that supported a government policy he opposed. The same moral tension led Bertrand Russell to be imprisoned for protesting British involvement in World War I and later in the Vietnam War. From the early abolitionists in England to Mahatma Gandhi's nonviolent resistance in India, from the suffragettes' campaign for women's rights to Rosa Parks's defiance in segregated Alabama, from Nelson Mandela's struggle against apartheid to climate scientist James Hansen's arrest for opposing a tar sands pipeline, the progress of civil rights and social justice has often been driven by acts of dissent and civil disobedience.

Many of these moral pioneers have spent years behind bars. Others have paid with their lives. Among them are Steve Biko, a fierce anti-apartheid activist from South Africa beaten to death by the police while in custody; Anna Politkovskaya, a Russian journalist who was murdered for exposing war crimes in Chechnya; and Rachel Corrie, who was crushed to death by an Israeli bulldozer while trying to stop the demolition of a Palestinian home in Rafah.

Since October 7, numerous individuals have engaged in acts of resistance, refusing to remain silent in the face of the slaughter in Palestine. A growing number of dissenters have spoken out and taken personal risks to refuse complicity. One notable form of resistance has come from within: Israeli citizens refusing to enlist in the military. Ben Arad is among them. He declared his refusal to serve in the Israeli Army moments before reporting to the recruitment center near Tel Aviv on April 1, 2024. "I am willing to pay a price for my principles," he said. "Since the war began, I have understood that I have an obligation to make my voice heard and call for an end to the cycle of violence." Arad was sentenced to 20 days in a military prison and became the third Israeli teenager to publicly refuse conscription for political reasons since October 7, 2023. The first was Tal Mitnick, who served 105 days; the second was Sofia Orr, who served 40 days (Reiff, 2024a).

Their actions may have inspired others. Three more 18-year-olds, Yuval Moav, Oryan Mueller and Itamar Greenberg, openly declared their refusal to enlist. Each was subsequently tried and sentenced to military prison. In a public statement to Palestinians, Moav said: "In my simple act, I want to stand in solidarity with you." He also added: "I also recognize that I do not represent the majority opinion in my society. But in my action, I hope to raise the voice of those of us who are waiting for the day when we can build a common future [and] a society based on peace and equality, not occupation and apartheid" (Oren, 2024).

In March 2025, 18-year-old Ella Keidar Greenberg was sentenced to thirty days in an Israeli military prison for refusing to enlist in the army. The first openly transgender conscientious objector in over a decade, Keidar cited her refusal as a stand against both the occupation and Israel's assault on Gaza. "Faced with a reality of mass extermination, of systematic neglect, of trampling on rights, of war—the imperative is refusal," she said. Before being taken to prison, she made a public statement saying: "When our grandchildren ask us what we did during the Gaza genocide—if we gave up or if we put up a fight—how would you rather answer? I know what I'll answer: that I chose to resist. This is why I am refusing" (Ziv, 2025).

Outside Israel, some Jewish organizations and individuals have been at the forefront of global dissent. From Washington, D.C. to Glasgow, from London to Paris, from Rome to Barcelona, Jewish demonstrators have stood shoulder to shoulder with other activists, human rights defenders, and concerned citizens in denouncing the genocide. In the United States, the non-governmental organization *Jewish Voice for Peace* has spearheaded a wave of protests and acts of civil disobedience, joined by a network of Jewish groups across Europe. In Italy, over two hundred Jewish citizens signed a letter titled "Jews of Italy Say NO to Ethnic Cleansing." The collective message of all these forms of resistance, "Not in my name", is both a moral indictment of the Israeli government and a call to disentangle Jewish identity from Israeli state violence. The message underlines that to be Jewish is not to endorse occupation or war. And to stand against genocide is not betrayal. In fact, it is the fulfillment of Jewish ethical tradition at its most courageous.

1.2 *US Elections 2024: "I Can't Vote for Genocide"*

In November 2024, Donald Trump won the U.S. presidential election. Again. The anguish that followed was immediate and justified. Across the country and beyond, voters, commentators, and ordinary citizens voiced deep concern about the resurgence of a political figure whose initial presidency had destabilized democratic institutions, undermined science, and normalized authoritarian rhetoric. Within weeks, Trump's new administration began an aggressive assault on civil liberties, climate policy, public institutions, and free expression. He had once again assumed control of the most powerful office on Earth, this time with fewer constraints and opposition.

Just days before the election, I appeared on the Italian national TV program *Leonardo: Scienza e Ambiente* to discuss the psychology of the U.S. electoral landscape. Alongside well-known factors such as political polarization, partisan animosity, and the dehumanization of political opponents, I examined a growing disillusionment of American voters with the two-party system. I also tried, without success, to mention another factor that would later contribute to Kamala Harris's debacle: the genocide in Gaza (RAI, 2024).

To be clear, the gap in voter support between Trump and Kamala Harris was significant. While numerous political, economic, and social factors influenced the erosion of Democratic support among Latino voters and working-class Americans, Gaza emerged as an additional breaking point. Harris's complicity in the slaughter of civilians in Palestine alienated voters who could no longer reconcile their political loyalty with their conscience. Many people who had previously voted Democratic must have had this thought in their mind: "I can't vote for genocide."

Kamala Harris's refusal to stop the flow of arms to Israel symbolized a broader failure: her inability to appeal to the most progressive members of the Democratic Party. On Gaza, as on immigration, the economy, and other key issues, her campaign seemed more focused on persuading elites and moderate Republicans than on energizing the Democratic base or mobilizing passionate, young voters. As Bernie Sanders stated: "It should come as no great surprise that a Democratic Party which has abandoned working class people would find that the working class has abandoned them." (Walker, 2024).

A poll by the *Institute for Middle East Understanding Policy Project/YouGov* found that Gaza was cited as the main reason not to vote in 2024 by a significant share of Biden 2020 voters. About 29% of these voters said "ending Israeli violence in Gaza" was the most important issue (IMEU Project, 2025). Although no single factor can fully explain the outcome of the U.S. election, the results in Arizona, Michigan, Wisconsin, and Pennsylvania showed that complicity in the Gaza genocide helped persuade Biden voters not to vote for Harris (Rascius, 2025). This finding aligns with a pre-election poll conducted in three pivotal states (Georgia, Pennsylvania, and Michigan), which revealed that the Gaza humanitarian crisis was a primary political concern for the majority of Muslim voters (61%) (Mogahed, 2024).

Of course, polling has its flaws, particularly the challenge of self-reporting bias. However, the hard data from the 2024 election speaks for itself. Kamala Harris received approximately 75 million votes, which is far fewer than the 81 million votes Joe Biden received in 2020. The loss was not only in the total number of votes, but also in where those votes were lost. Donald Trump's victory hinged on three swing states: Pennsylvania (19 electoral votes), Michigan (15), and Wisconsin (10). Had Harris won in those states, she would have reached the 270 electoral votes needed to be elected U.S. President. Instead, she stalled at 226.

Michigan, home to one of the largest Muslim populations in the United States, felt the effects of the Democrats' complicity in the Gaza genocide. In cities like Dearborn, Harris lost more than half of the votes Biden received in 2020 (Warikoo, 2024). Congresswoman Rashida Tlaib, a Palestinian American known for unapologetically condemning Israel's war crimes, including holding up signs reading "War Criminal" and "Guilty of Genocide" during Prime Minister Netanyahu's congressional address, won 69.7% of the vote in her district which includes cities such as Detroit and Dearborn.

In Wisconsin, the Gaza genocide also significantly influenced voting behavior. A protest campaign urging voters to submit "uncommitted" ballots garnered more than 47,000 votes in the Democratic primary. This symbolic gesture became a reality in November when many Democrats abstained from supporting Harris (Rama, 2024).

In Pennsylvania, dissatisfaction among Hispanic voters over the administration's economic policies and performance played the most important role. However, the defection of disillusioned progressives and Muslim voters also contributed to the margin that handed Trump the state (Lacy, 2024).

Some Democrats have expressed shock and outrage at Trump's return to power. But they have only themselves to blame. Trump did not win the 2024 election. It is the Democratic Party that lost it. The Party chose complicity and silence over moral principles. They chose political suicide over stopping a genocide.

As a *Slate* article bluntly summarized, voters did not defect due to apathy or economic grievances alone; they also defected because they refused to support the Democratic Party. They could not, in good conscience, endorse a party whose policies helped enable the mass killing of Palestinian civilians. Many stayed home. Others voted for Jill Stein, the only candidate who unequivocally opposed the atrocities in Gaza and the continued flow of weapons to Israel. "Yes … they will vote their conscience, even if it means handing the White House to Donald Trump," the article said (Ismail, 2024).

In 1992, Bill Clinton won the presidency with a slogan that captured the political moment: "It's the economy, stupid." In 2024, the refrain that caused Harris to lose may be: "It is not just the economy; it is also the genocide, baby."

2   Medics in Gaza's Last Hospitals

December 27, 2024. Amid the rubble of destroyed buildings, a man in a white coat approaches two massive tanks blocking the road. He is Dr. Hussam Abu Safiya, a pediatrician and neonatologist who is also the director of the Kamal Adwan Hospital in northern Gaza. Meanwhile, patients, medical staff and families are forced to evacuate. Abu Safiya is then arrested. His crime? Saving lives, even if it meant sacrificing his own.

Kamal Adwan Hospital was one of the last major health facilities still operating in northern Gaza. Israeli forces carried out yet another "military operation," setting fire to parts of the hospital and severely damaging key areas, including the laboratory, surgical unit and medical storage area. Dr. Safiya and several other health workers were arrested, and critically ill patients were transferred to a health center that lacked the necessary equipment and supplies for proper treatment.

The Israeli army repeatedly warned Dr. Safiya and other health workers to leave the hospital before bombing it. Faced with the risk of death, they refused. They insisted that they were committed to the humanitarian oath they took when they began their medical careers, vowing to provide care to those in need. "We will leave when the last Palestinian leaves northern Gaza," Abu Safiya made clear (Hajjaj, 2024).

According to the *World Health Organization* (*WHO*), the bombing of Kamal Adwan Hospital was the culmination of a sustained wave of assaults aimed at systematically dismantling the healthcare system. For the previous two months, hospitals and medical personnel in the region had been regularly attacked. Meanwhile, requests for the deployment of international emergency medical teams were denied (WHO, 2024).

During almost two years of genocide, the WHO has repeatedly called for the protection of health workers and hospitals in accordance with international humanitarian law. However, these calls have gone unanswered. Attacks and violence have continued in flagrant disregard of the Geneva and Hague Conventions, the Rome Statute, and international humanitarian law. Countless open letters signed by hundreds of thousands of citizens, doctors, nurses, and other professionals have gone unanswered. As Dr. Safiya explained: "We have been asking the world for international protection of the health system. These are laws established by the Geneva Conventions, which stipulate the protection of the health system." He added, "Where are these laws? What sin have we committed in this hospital to be bombed and killed like this?" (Hajjaj, 2024).

During the pandemic, doctors and nurses on the front lines were hailed as heroes. However, when it comes to heroism, the deeds of Palestinian and volunteer doctors and other medical personnel in Gaza are unmatched. Those who witnessed the atrocities of the genocide firsthand, while trying to save lives under constant bombardment, are epitomes of generosity and professionalism (Mahase, 2023).

According to the *WHO*, between October 2023 and May 2025, the Israeli military has carried out 720 attacks on healthcare targets in the Gaza Strip, including 125 health facilities, 34 hospitals and 186 ambulances (WHO, 2025; UNOCHA, 2025c). More than 1,700 healthcare workers have been killed. This, of course, implies another record for Israel: the highest number of health professionals killed in a war zone in recent history.

According to Mohammed Al-Sharif, who wrote an editorial while the Kamal Adwan hospital was being bombed day and night: "The sounds of explosions and bullets do not stop. The hospital is in a constant state of panic. With each new explosion or round of gunfire, patients flee from one wing of the hospital to another, crowding into the narrow hospital corridors to sleep like sardines, hoping to be safe" (Hajjaj, 2024).

In April 2024, Amber Alayyan, a pediatrician with *Doctors Without Borders*, stated during a press conference at the United Nations that "doctors are faced with the terrible decision of having to intubate and amputate children and adults in emergency rooms without anesthesia" (Doctors Without Borders, 2024b).

Professor Nick Maynard, an Oxford surgeon who has traveled to Gaza since 2010, witnessed with his own eyes an Israeli army attack on the intensive care unit while operating on a patient with abdominal and chest injuries from a bomb. "They have been deliberately targeting health workers, health vehicles and health buildings since October 7," Maynard said (Abdul, 2024).

Do you remember the heated media controversies that followed the deadly explosion in the courtyard of Al-Ahli Arab Hospital in Gaza City on October 17, 2023? The Palestinian Ministry of Health reported that 471 people lost their lives and 342 were injured in the attack.

Many mainstream media outlets including *ABC*, (Deliso, 2023) *The Wall Street Journal*, (Wall Street Journal, 2023b), *CNN* and *BBC*, (Forensic Architecture, 2024b) suggested that the explosion was likely caused by a misfired Hamas rocket from Gaza. *The Economist* titled the story "What is Palestinian Islamic Jihad? Israel blames the group for a deadly explosion at a hospital in Gaza" (The Economist, 2023).

By November 2023, however, investigations by *Channel 4 News, Al Jazeera*, and *Earshot* disputed these claims. A first report from *Forensic Architecture* in October 2023 as well as a subsequent investigation in 2024 that included 3D tracking of the alleged rocket volley and testimony from an on-site doctor, undermined the misfire theory (Forensic Architecture, 2024c). The massacre was caused by what seemed to be the most obvious suspect: the Israeli army.

At the time, many Western intellectuals and media pundits also insisted that Israel would never target hospitals, despite ample evidence to the contrary. According to the *WHO*, from 2019 to 2021, there were 563 attacks against health care facilities, with 288 in the West Bank (of which 93 were in east Jerusalem) and 275 in the Gaza Strip (WHO, 2022).

Of course, these assaults are not comparable to those perpetrated after October 7. Here is the chronology of Israel's attacks developed by *Doctors Without Borders* until the end of 2024 and updated with some of the attacks conducted in 2025 (Doctors Without Borders, 2025):

– Al-Ahli Arab Hospital: October 17, 2023; April 13, 2025
– Al-Aqsa Hospital: January 6, 2024; March 31, 2024; July 22, 2024; August 4, 2024; August 25, 2024; September 5, 2024; October 7, 2024; October 14, 2024; November 9, 2024
– Al-Awda Hospital: October 11, 2023; October 13, 2023; November 21, 2023; December 1, 2023; December 5, 2023; December 12, 2023; December 17, 2023; May 23, 2024
– Al-Emirati Hospital (Rafah): March 2, 2024
– Al Shaboura Clinic: March 27, 2024
– Al-Shifa Hospital: November 3, 2023; November 10, 2023; November 15, 2023; March 24, 2024; April 1, 2024

- Indonesian Hospital/Rafah Indonesian Field Hospital: October 7, 2023; May 12, 2024; May 18, 2025
- Kamal Adwan Hospital: October 26, 2024
- Khalil Suleiman Hospital (Jenin): December 14, 2023; March 12, 2024; September 5, 2024
- Doctors Without Borders Gaza Clinic: October 10, 2023
- Doctors Without Borders Premises near Al-Shifa: November 14, 2023
- Doctors Without Borders Evacuation Convoy: November 18, 2023
- Doctors Without Borders Convoy: November 22, 2023; November 24, 2023
- Doctors Without Borders Lotus Shelter, Khan Younis: January 8, 2024
- Doctors Without Borders Shelter, Al Mawasi: February 20, 2024
- Doctors Without Borders Clinic: April 1, 2024; June 25, 2024; July 8, 2024; November 13, 2024
- Doctors Without Borders Stabilization Point, Tulkarem/Nur Shams: May 6, 2024
- Doctors Without Borders Healthcare Center, Al Mawasi: May 30, 2024
- Doctors Without Borders Healthcare Facility, Al Mawasi: December 17, 2024
- Nasser Hospital: October 7, 2023; January 16, 2024; January 22, 2024; February 15, 2024; December 17, 2023; March 23, 2025
- Tal Al Sultan Stabilization Point: May 28, 2024
- Tulkarem and Nur Shams Camps (Medical Points): April 21, 2024; May 6, 2024
- Turkish Hospital (Tubas): December 3, 2024
- Turkish-Palestinian Friendship Hospital: October 30, 2023; March 21, 2025
- Gaza European Hospital (Khan Younis): May 13, 2025

As noted in an article published in the academic journal *South African Journal of Bioethics and Law*, "Sanctuaries of humanity have been transformed into corridors of horror" (Mahomed, 2023). Nada Abu Alrub, an Australian doctor who has been volunteering at Al-Shifa Hospital in Gaza City, told Al Jazeera, "Hospitals have turned into graveyards." She added: "You don't know whose hand is this and whose leg is this—it's kind of like a horror movie."

There are countless heartbreaking stories of medical professionals who were killed in Gaza, as well as testimonies about their extraordinary dedication. One such story is that of Dr. Hammam Alloh, a 36-year-old Palestinian nephrologist at Al-Shifa Hospital. On October 31, 2023, an Israeli artillery shell struck the home of his wife, killing Dr. Alloh, his father, his brother-in-law, and his father-in-law.

Only two weeks before his death, Dr. Alloh explained in an interview with Democracy Now! why he had refused to evacuate to southern Gaza, as the

Israeli military had ordered. He replied: "If I leave, who will treat my patients? We are not animals. We have a right to proper medical care. Do you think I spent fourteen years in medical school and postgraduate studies thinking about my life and not my patients?" (Democracy Now!, 2023b).

3   "Stop the Genocide!"

On June 3, 2024, the Italian newspaper *La Repubblica* published an article by the British academic and writer Denis MacEoin, titled *"Dear Students, Israel Is Not a Regime."* The piece appeared in the cultural section of the newspaper at a time when waves of student-led protests erupted across Western universities in solidarity with Palestinians. However, the editors failed to disclose, or perhaps even to notice, that the letter was written in 2011. Although Denis MacEoin passed away in 2022, the article was presented as if it were a direct and timely response to the current student protest movement (MacEoin, 2024).

This oversight raises questions about journalistic integrity. It reveals how bias can cloud professionalism or how desperate the mainstream Western media had become to discredit students who dared to speak out against the daily massacres in Gaza. Moreover, does it really matter to the victims of a genocide whether the perpetrators claim to be a democracy? Is genocide more acceptable just because it is committed by an elected government? In October 2024, I participated in the 38th Venice Half Marathon. As is tradition, each runner could have their name printed on the official T-shirt. The night before the race, I went to the T-shirt booth at the marathon expo. "What first and last name should I put?" asked the attendant, ready to print it on the shirt. "Stop the genocide," I replied. He widened his eyes: "Excuse me?" "Stop is the first name and The Genocide is the last name," I joked.

3.1   *Students Teaching the World a Lesson*

In early May 2024, CBS *News Sunday Morning* aired a segment featuring a new exhibition in New York City titled *The Nova Music Festival Exhibition: October 7th/06:29am, The Moment Music Stopped.* The exhibit was created to memorialize the victims of the *Tribe of Nova* music festival transformed into a massacre during the Hamas-led attacks of October 7, 2023. Through curated artifacts and haunting video footage, the exhibition captured the pain and horror of what should have been a celebration of music and life, now remembered as a crime against humanity (CBS Sunday Morning, 2024).

Yet, the focus of the *CBS program* took a sharp turn. The host used the moment of commemoration for the victims to comment on the student-led protests erupting across U.S. university campuses at the time. Referring to the protests in solidarity with Gaza, she dismissed them with disdain and questioned their morality. She said, "Mutual empathy? A casualty." She probably meant to say that the students' display of compassion for Gaza demonstrated a lack of empathy for the Israeli victims of October 7, 2023.

Unwittingly, the *CBS* host offered a textbook example of how selective empathy works: grief is to be reserved only for some, our in-group, and others, in the out-group, are not worthy of mourning.

Scooter Braun, the artist who brought the exhibit to New York, however, did not seem to be on the same page. A different type of empathy emerged from his words when he said: "Music should be a safe place ... Have empathy in your heart for both sides" (CBS Sunday Morning, 2024). In that simple plea, Braun gave expression to what the student protesters had been demanding all along: that empathy should not be exclusive, but extended to all victims, regardless of nationality, religion, or politics.

The students, often caricatured or condemned in mainstream media, were in fact demonstrating the very principles democratic societies claim to cherish: the refusal to accept a hierarchy of suffering. Far from being naïve or radical, their protests served as a powerful civic lesson, one that reminded the world that empathy, if it is to mean anything at all, must extend to every human being.

Driven by moral clarity and a refusal to look away, their protests grew. They set up encampments, held teach-ins, and organized petitions. From the U.S. to Europe, from Canada to Australia, a global student movement emerged. As *The Guardian* wrote, it became "the most significant student movement since the anti-Vietnam campus protests of the late 1960s" (Helmore, 2024). And what were their demands? There were many: some called for divestment from Israel, others demanded an end to academic and institutional collaboration. At its core, however, the movement was about protesting the moral issue of our time.

Some of these students have paid a high price for standing in solidarity with Gaza. Since the university encampments began in the U.S., over 3,100 protesters have been arrested on more than 60 campuses. Mass arrests have also occurred in some European countries, including the United Kingdom and the Netherlands. Some of these students have faced suspension or expulsion from their academic programs. They have put their entire future on the line for Gaza.

According to an article reprinted from *The Conversation* and published in *Scientific American*, major media outlets have distorted both the goals and actions of the student protests. Rather than addressing the moral urgency behind the movement, the coverage has fixated on arrests and sensationalism (Brown & The Conversation, 2024). A study found that *The Globe and Mail*, a leading national newspaper in Canada, unfairly portrayed the student protesters in an overwhelmingly negative light. Coverage of the *McGill University* encampment frequently emphasized alleged anti-Semitism and hate speech. Discussions of security repeatedly framed the protesters' impact on campus. Terms such as "violent" and "violence" appeared 69 times in forty articles to describe the protesters (Innes, 2024). An episode of *MSNBC*'s *Morning Joe* also accused the student movements of being supporters of Hamas motivated by hatred toward all Jews (Decker, 2024).

But were these students violent as Western mainstream media insisted?

According to research by the *Armed Conflict Location and Event Data Project*, an independent nonprofit organization that investigates political violence and protests around the world, 97% of demonstrations on U.S. campuses were peaceful. Of the 553 demonstrations analyzed nationwide between April 18 and May 3, 2024, fewer than 20 resulted in serious interpersonal violence or property damage (Beckett, 2024). As for the accusations of anti-Semitism, major media outlets relied on anecdotal stories or the emotional reactions of a few students who felt "unsafe" on campus. Yet, they have provided no evidence that these students' attitudes were anti-Semitic. In fact, most students' protests were overwhelmingly peaceful, and focused on justice and human rights, not hatred.

Students at my own university too, including some from my courses, displayed remarkable courage when they set up camp at the historic Palazzo Bo' to urge the University of Padova to take action against the genocide. During an event in Vicenza, one of them expressed his disappointment that the University's Academic Senate, after having condemned the atrocities of Hamas in clear and unequivocal terms, used much weaker language to describe the crimes committed by Israel. At one point, the Academic Senate's document stated that Palestinians had suffered "unacceptable human losses and discomfort." As observed by the student, "discomfort might be used to describe a delayed train, but not a genocide."

3.2   *The Academic Witch-Hunt*

"Only Israel can murder around 300 children in the span of a few weeks and insists that it is the victim." This is a Twitter message, from September 2014, that cost University of Illinois professor Steven Salaita his job (Guarino, 2014).

He was accused of being "critical of Israel." Yet, according to the *United Nations*, between July 8 and August 27, 2014, during *Operation Protective Edge*, the Israeli army killed 2,104 Palestinians, including 1,462 civilians. Of them, 495 were children and 253 were women (UNOCHA, 2014). In other words, evidence proves that Professor Salaita's statement was accurate, but conservative.

According to the media outlet *News-Gazette*, the decision to suspend Salaita came after some donors threatened to withdraw their financial support if the university did not act to punish or suspend the professor. "I am deeply conflicted by my decision to reconsider any support for the business school" said one of them. "However, as a Jew and lover of Israel, I see no other way to make my voice heard than to take this action," the message continued (Des Garennes, 2025).

Prof. Salaita became just another victim of a global, concerted, well-funded, and organized effort to silence those who show empathy for Palestinians. This incident also highlights a blatant violation of academic freedom within a general trend affecting academia over the past half-century. The increasing privatization of universities and their dependence on private funding and alumni donations from wealthy donors has come at a price: the liberty of academics to study topics that challenge some power and economic interests.

Academic freedom in the U.S. has been under attack for a long time. One of the most striking examples is *DePaul University*'s denial of tenure to Norman Finkelstein, despite his impressive international academic recognition. Finkelstein, a political scientist with a Ph.D. from *Princeton University* and the subject of the documentary *American Radical*, is widely recognized as a leading analyst of Palestinian history and politics (Ridgen & Rossier, 2009). In an interview with the media outlet *Democracy Now!* he exposed the numerous factual inaccuracies in *The Case for Israel*, a book by *Harvard University* professor Alan Dershowitz, calling it "a hoax" (Democracy Now!, 2003). A voice like no other, Finkelstein has contributed more than anyone else to our understanding of the topic. Since being denied tenure, he has never been granted a proper academic position, despite his unique analytical rigor, clarity of thought and widely praised scholarship.

If academic freedom was already under threat before October 7, the events that followed that day have deepened the crisis. Since then, there has been a clearly observable campaign to silence dissent, amounting to a modern-day witch-hunt against academics and scholars who have spoken out about the genocide. Many have faced suspension, dismissal, or public vilification. In some cases, scholars and prominent professionals have even been arrested or criminally investigated for expressing their views.

DePaul University made headlines once again for firing adjunct professor Anne D'Aquino in the middle of her first quarter teaching *Health 194: Human Pathogens and Defense* because some students had expressed concerns about an assignment focusing on Palestine that included the terms "genocide" and "ethnic cleansing." The assignment asked students to explain the impact of genocide and ethnic cleansing on the health and biology of those affected (Quinn, 2024).

That same month, Sang Hea Kil, a professor in the *Department of Justice Studies* at *San Jose State University*, was suspended. She was accused of engaging "in behavior that disrupted the university's business operations and encouraged students to do the same." This prompted the resignation of one of her colleagues, Professor Rochelle McLaughlin who had taught in the *Department of Occupational Therapy* since 2004. In her letter of resignation McLaughlin wrote: "I have been teaching here for the past twenty years and I can no longer ethically or morally consent to use my labor in any way that supports an institution that is explicitly manufacturing consent for the unrestrained, illegal, 'apocalyptic' ongoing series of 'calculated' genocides to destroy the people of Palestine" (R. McLaughlin, 2024).

Benoît Huou, a math teacher at the *Toulouse School of Economics* in France, was suspended after a student recorded him making the following comment in a class: "In my lifetime, I'm 35 years old, I've never witnessed such a massacre, such a one-sided war." Huou's statement was evidence-based. He quoted a study published in *The Lancet* that estimated that the Israeli onslaught had caused approximately 180,000 deaths in Palestine since October 7 (Le Nevé, 2024).

That same month, Maura Finkelstein, a tenured professor at *Muhlenberg College* in Pennsylvania, was fired after sharing an Instagram post by Palestinian poet Remi Kanazi that called for rejecting Zionist ideology and its supporters. The post read: "Don't cower before Zionists. Shame them. Do not welcome them in your rooms. Why should these genocide-loving fascists be treated any differently than any other outright racist?" (Sainato, 2024).

Steven Thrasher, a journalism professor at *Northwestern University*, was suspended after being charged by police over the summer for trying to protect students from arrest at their protest encampment (Sainato, 2024).

Nancy Fraser, a Jewish American professor of philosophy and politics at the *New School for Social Research* in New York, was disinvited from accepting a prestigious professorship at the *University of Cologne* after signing a letter expressing solidarity with Palestinians and condemning the killings in Gaza by Israeli forces (Connolly, 2024).

Linguistics professor Michel DeGraff of the *Massachusetts Institute of Technology (MIT)*, submitted a proposal to teach a course on the Israeli–Palestinian conflict. In the following months, he became embroiled in an ongoing dispute with his department chair and colleagues. Meanwhile, the *MIT* dean issued a formal reprimand and withheld a salary increase. DeGraff's colleagues argued that the course, which would have examined the language used in discussions of the Israeli–Palestinian conflict, did not fit within the department's curriculum and that DeGraff lacked the necessary expertise to teach it (Friedman, 2024).

Joseph Daher, a Swiss-Syrian academic at the *University of Lausanne*, discovered that his course, *History of International Relations Post-1945*, with a special focus on the Middle East, had been abruptly canceled. This left the approximately sixty students enrolled in the course confused and without an explanation from the administration. Although contract renewals for future courses are usually routine, this cancellation effectively ended Daher's long-standing relationship with the institution. Daher believed the university's decision was due to his outspoken support for Palestinian rights (McLoughlin, 2025).

Amin Husain, an adjunct professor at *New York University*, was suspended for denying reports that Hamas militants are "rapists" and "behead babies" during a teach-in. He defined those claims as "not true." A petition launched on October 17 has gathered over 6,700 signatures and called for Husain's dismissal, alleging that he had spread hate speech targeting Jewish people (Zipfel & Nehme, 2024).

Nadera Shalhoub-Kevorkian, a professor and dual American Israeli citizen who has taught at *Hebrew University* for nearly three decades and authored over 100 articles and books, was arrested and suspended for comments accusing Israel of genocide in Gaza and casting doubt on reports of sexual violence by Hamas. Professor Shalhoub-Kevorkian's arrest prompted one of her colleagues, Yuri Pines, a professor of Chinese Studies at the same university, to announce his resignation in solidarity. Pines wrote a letter to the university stating that, "The content in the vile letter about the dismissal of Nadera Shalhoub-Kevorkian" surprised him. "I never thought that the Hebrew University was a Zionist institution: I saw it as an academic institution, in which Zionists and non-Zionists, as well as anti-Zionist people like myself, could work." The letter added: "I thought that the university was led by people who are rational enough to understand that the issue of whether Israel is committing genocide in Gaza belongs to the field where students and lecturers can freely express their opinion" (Osman, 2024).

Haim Bresheeth, a retired Jewish professor, child of Holocaust survivors, and founder of the *Jewish Network for Palestine*, was arrested under a British anti-terrorism law after speaking at a pro-Palestinian protest in London. Despite being an advanced cancer patient with a heart condition, he spent time in prison without being told what his crime was (Seidman, 2024). Another notable example is Fiona Godlee, former editor-in-chief of the British Medical Journal, who was arrested in London for peacefully protesting genocide.

Irene Khan, the *UN Special Rapporteur on Freedom of Expression and Opinion*, observed that the Gaza crisis has become "a global crisis of the freedom of expression," affecting "academic freedom in the United States (and other Western democracies)." She added that the crisis infringes on "people's rights to protest a war and occupation" (United Nations, 2024c).

Not even George Orwell when he wrote one of his best novels, *1984*, could have imagined a society where opposing genocide makes you a criminal.

4 Just for a Cause

"Imagine living in a world where you can be persecuted for saying, 'don't kill children,' because you might hurt the feelings of the murderer." This slogan, spotted during a peace demonstration, needs no further explanation. It reads like a description of a dystopian society, the kind you would expect in a fantasy novel or in a movie about a world gone mad. But it is not fiction. It speaks to the current reality in countries that many of us once believed to be strongholds of freedom, democracy, and human rights.

And yet Gaza has also become a crossroads of conscience. The tragedy has forced a generation of intellectuals to take a moral stand. Many who remained silent clung to ideas of neutrality, claiming that their work required objectivity and detachment. Others displayed the full force of their universal sense of compassion. In this overwhelming tragedy and time of darkness, moments of light appeared in unexpected places and people, offering flashes of moral courage. Despite censorship, suspensions, dismissals, arrests, and attempts to silence dissent, a new generation of activists, students and concerned citizens has tried to awaken a society that has normalized a genocide.

At times, some of those who challenged power, often risking their status, careers, or even their lives in defense of a "just cause," have felt a deep sense of despair and hopelessness. For months, their massive efforts, activism, writing, petition signing, public meetings, and protests, seemed to achieve little. "It was all for nothing," some said. Despite all efforts, Israel's military operations continued. Nothing seemed to stop the daily massacres.

Amid that despair, an important question arises: why would a globally organized, massively funded public communication system invest so much effort and resources into silencing and persecuting students and intellectuals who are protesting the genocide in Gaza?

The answer came, unintentionally, from Alex Karp, the CEO of *Palantir Technologies*, a private military contractor that uses artificial intelligence systems to carry out military operations in Gaza. In response to the mobilization of students at U.S. universities and around the world, Karp observed: "We somehow think that these things that are happening on college campuses are a sideshow. No, they are the show. If we lose the intellectual debate, you will not be able to deploy an army in the West, ever" (Mazza, 2024).

Karp is right. Western public opinion and taxpayer money play an important role in enabling the Israeli war machine to continue its activities. Karp recognized the danger of losing the public relations battle. And plenty of evidence suggests that Israel is losing it.

In May 2025, after more than eighteen months of relentless devastation in Gaza, something changed. Major British media outlets finally published clear-cut editorials condemning Israel. This shift marked a significant U-turn in mainstream discourse, particularly among publications that have historically been cautious and aligned with Israel's narratives.

The *Financial Times* published a striking editorial titled *The West's Shameful Silence on Gaza*. The article sharply criticized the U.S. and European governments for "issuing barely a word of condemnation" toward their close ally. It concluded with a blunt moral indictment: "They should be ashamed of their silence and stop enabling Netanyahu to act with impunity" (The Financial Times, 2025). Shortly after, *The Economist* followed suit with an editorial titled *The War in Gaza Must End*. The article called for urgent international pressure, specifically naming U.S. President Donald Trump to compel the Netanyahu government to halt its military offensive (The Economist, 2025).

Weeks later, as Gaza endured yet another blockade with no food or humanitarian aid for nearly a month, Western governments finally took some first measures. The EU announced a review of its trade agreements with Israel based on human rights grounds. The United Kingdom suspended trade talks and summoned its ambassador regarding the blockade of Gaza. In September 2025, Spain began taking concrete action. The country imposed a total arms embargo, recognized the genocide, imposed diplomatic sanctions, and initiated steps to suspend trade agreements. As Spanish Prime Minister Pedro Sánchez pointed out: "We do not trade with a genocidal state." A tipping point?

Perhaps. At the time of writing, the future remains uncertain. The humanitarian catastrophe continues, and Israel remains unpunished. While welcome, recent political actions by Western governments are far too little, far too late.

And yet, the wall of silence and complicity is beginning to show some cracks. Across Italy, Europe, and the world, millions have flooded the streets. Students, workers, doctors, teachers, and artists are united by one demand: Stop the genocide! From Rome to Madrid, from London to Johannesburg, from New York to Jakarta, a moral awakening is unfolding.

Those once silent are beginning to speak. Those once afraid are beginning to act. The narrative has shifted. Israel is losing its public relations war, and its true intentions are now recognized by most for what they are: genocidal.

They tried to erase Gaza from the world, but Gaza gave birth to a new world.

Activism is making a difference. It is not all in vain. Quite the contrary.

And even if it were, one does not fight for a just cause because one expects to win. One fights for a just cause because it is just.

4.1   *Poetic Injustice*

Those who despair about the seeming inconsequence of activism should know that social change has always occurred because small groups of determined individuals decided to make a difference. History shows that change can happen even when our efforts seem to have little impact. It is the determination of impassioned activists that catalyzes significant societal advancements, redefining the course of societies and the world at large. As Schopenhauer once said, "All truth passes through three stages. First, it is ridiculed. Second, it is violently opposed. Third, it is accepted as self-evident" (Chu, 2016).

Bertrand Russell asserted that there are hidden but profound rewards for the soul of those who seek a better world. "Those who live nobly" he wrote, "need not fear that they have lived in vain. Something radiates their life, some light that shows the way to their friends, to their neighbors ... I find many people nowadays oppressed by a sense of helplessness, with the feeling that in the vastness of modern societies there is nothing important that individuals can do. This is a mistake." Then he added: "The individual, if he is full of love for humanity ... can do a lot: each of us can broaden his mind, free his imagination, and spread his affection and benevolence. And it is those who do so that humanity ultimately venerates" (Russell, 1950).

Taking a stand for a just cause, of course, carries consequences. Activists and free thinkers who have exposed the ongoing genocide in Gaza have faced severe retaliation. Some have been defamed, threatened, suspended, imprisoned, and in some cases killed.

Before he was assassinated, Refaat Alareer, a poet, writer, activist, and professor of comparative literature at the *Islamic University of Gaza*, wrote his last poem, *If I Must Die*. Less than a month later, he and seven members of his family were killed when Israeli forces deliberately targeted his home in an airstrike.

Refaat Alareer's death was more than a personal tragedy. His death became a profound indictment of the injustices suffered by Gaza, and a reminder of the price of speaking truth to power.

Chris Hedges paid tribute to the poet in a powerful, imaginative letter, writing: "Dear Refaat, you have joined the ranks of martyred poets. The Spanish poet Federico Garcia Lorca. The Russian poet Osip Mandelstam. The Hungarian poet Miklós Radnóti, who wrote his last verses on a death march. The Chilean singer and poet Victor Jara. The black poet Henry Dumas, shot dead by the New York City police" (Hedges, 2024b).

> If I must die, you must live
> to tell my story, to sell my things
> to buy a piece of cloth and some strings,
> (make it white with a long tail)
> so that a child, somewhere in Gaza
> while looking heaven in the eye
> awaiting his dad who left in a blaze—
> and bid no one farewell, not even to his flesh, not even to himself—
> sees the kite, my kite you made, flying up above
> and thinks for a moment an angel is there
> bringing back love
> If I must die, let it bring hope, let it be a tale.
>
> REFAAT ALAREER

On the Brink

If you feel pain, you are alive. If you feel other people's pain, you are
a human being.
LEO TOLSTOY

⁘

Amid the ongoing tragedy in Gaza, countless people across the West and
around the world have experienced a parallel sense of anguish, an inner tur-
moil stirred by the images of violence and suffering. Though only an infinitesi-
mal fraction of what Palestinians have endured, these images have left lasting
effects in the hearts of those who care about justice and humanity. Anyone
with even a shred of moral conscience must have felt the weight of guilt. Guilt
for watching, for not doing enough, for feeling powerless in the face of Gaza's
agony.

In a conversation with Chris Hedges, Gabor Maté posed a haunting question: "How can we live this horror and still survive as human beings?"

Richard Horton, editor of *The Lancet*, expressed a similar sense of despair,
lamenting the international community's failure to act in defense of Palestine.
He wondered whether this inaction reflected "the erosion of compassion
for one another." But survey data do not support the hypothesis of a general
decline in compassion (Lenharo, 2023). In fact, the international community
has shown that it can respond swiftly and decisively to some humanitarian
crises, mobilizing support, aid, and resources when it chooses to.

The problem, then, may not be a decline of compassion, but a selective
manifestation of it. What enabled the atrocities in Gaza is not a global failure
of solidarity, but a form of humanitarianism that does not extend to all human
beings.

1　A Crisis of Selective Humanitarianism

More than anything else in recent history, Gaza has revealed that humanity is
in a state of moral crisis. The *Universal Declaration of Human Rights*, along with

decades of globalization and civic progress, has fostered the hope, or perhaps the illusion, that the world was progressing toward a shared set of universal values. However, rather than advancing egalitarian interests based on transnational cooperation and compassion, the political reality of nation-states remained entangled in territorial, ethnocentric, and nationalist frameworks.

International law, by design, demands that its principles, sanctions, regulations, and even the compassion behind humanitarian interventions be applied equally to all peoples and all nations. Applying these standards inconsistently breeds mistrust, fuels perceptions of injustice, and erodes the very legitimacy that prompted the creation of such laws in the first place. Agencies charged with responding to crises, international organizations, regional and national bodies, humanitarian NGOs, and civil society have too often acted inconsistently. Some responses to humanitarian emergencies, wars, and disasters have been bold and immediate. Others have been delayed, minimal, or entirely absent. This double standard has led scholars to coin the term "selective humanitarianism" (Crossley, 2020).

History is full of examples of selective humanitarianism. The international community has shown strong solidarity in certain crises; not only during the Russian invasion of Ukraine, but on other tragic events as well. When the Charlie Hebdo attack shook France, for example, the world responded within days with one voice: Je suis Charlie. That attack left 17 dead and 11 injured. Yet in Gaza, even 50,000 children killed or injured were not enough for the world to say: *Je suis Gaza.*

The stark disparity in how the international community met the needs of populations affected by violence undermined its credibility. To maintain their legitimacy, humanitarian interventions must be applied universally. Conversely, selective moral indignation and politically convenient compassion send a disturbing message: some lives are more valuable than others. This deepens the divide between nations and erodes faith in global institutions responsible for upholding human rights and dignity.

Even worse, this duplicity fosters disillusionment. When democratic institutions are seen as morally bankrupt, preaching noble values while practicing double standards, people often turn away from them. In this vacuum, demagogues and authoritarians thrive, exploiting the very hypocrisy they oppose. Between the devil and a saint pretending to be one, people prefer the former. They go for the original.

If humanitarian values are to retain their moral force, they must be upheld consistently, not applied on a selective basis. When Western nations invoke human rights selectively, they drain those principles of meaning.

Selective humanitarianism results from a cognitive prison that limits our empathy. This trait erects walls between "us" and "them," diminishing our capacity to feel the pain of those outside our tribe, our race, our nation, our religion. And unless we break free of that mental cage, true universality of human rights will remain hollow rhetoric.

2   Vital Crossroads

Beyond the unspeakable tragedy it has generated, the genocide in Gaza has had profound worldwide repercussions. It has strained international relations and further undermined the fragile framework of global cooperation, particularly between Western powers and the Global South. Though these blocs have long diverged in worldview, the Gaza crisis has further widened the gap.

Many developing nations' political leaders increasingly view the West's commitment to international law as selective: rigorously applied to weaker states while powerful allies are shielded from accountability. This perceived double standard fuels the belief that international institutions serve the partisan interests of the Global North, rather than upholding universal principles of justice. The resulting sense of unfairness fosters nationalism and egotistical interests. This happens at a critical time when the world is facing challenges that no single country can solve alone. The most urgent of these threats are climate change and the danger of nuclear war, crises that demand an unprecedented level of coordinated action worldwide.

2.1   *89 Seconds to Midnight*

Closer than ever. According to *The Bulletin of the Atomic Scientists*, humanity has never been in more danger of catastrophe. The *Doomsday Clock*, as it is called, stands at just 89 seconds to midnight. Since 1947, the clock has measured our proximity to global disaster. Initially set at seven minutes to midnight, the clock now reflects a world teetering on the edge, mainly due to the climate crisis and the risk of nuclear war (Bulletin of the Atomic Scientists, 2025).

The existential danger posed by climate change is scientifically plausible and increasingly urgent. The *Lancet Countdown 2021 Report* described the climate change crisis as a "code red for humanity," identifying it as the most significant global health threat of the century (Romanello et al., 2021). Rising global temperatures, ecological disruptions, and extreme weather events are already causing widespread harm, including serious health consequences (IPCC, 2023).

Despite the propaganda from the fossil fuel industry and corporate libertarian think tanks, over 95% of climate scientists agree that climate change is real and is primarily driven by human activity. Interestingly, the consensus grows stronger as scientists' expertise in climate science increases (Cook et al., 2016).

Although climate change is often portrayed as an environmental crisis, it is much more than that. It is a global societal issue that threatens the very foundations of human civilization and our future survival as a species (Kemp et al., 2022; Willcock et al., 2023). Surprisingly or not, the most concerning consequences of climate change have been neglected in the scientific literature. As noted in an article by Huggel and colleagues, the academic community must now move beyond isolated studies of environmental and health impacts to clearly define and address the existential risks posed by the climate crisis (Huggel et al., 2022).

Warnings from climate scientists are starker and darker than ever. "There is ample evidence that climate change could become catastrophic," says an article published in the *Proceedings of the National Academy of Sciences* (Kemp et al., 2022). A study published in *Scientific Reports* estimated the probability of a catastrophic outcome at about 10% (Bologna & Aquino, 2020). Another analysis suggested a one-in-twenty probability of "catastrophic or existential consequences." The authors illustrated this with a sobering analogy: "It's like taking a flight with a one in twenty chance that the plane ... will crash" (Xu & Ramanathan, 2017). They also added: "We would never get on a plane with a one in twenty chance of crashing, so why are we willing to send our children and grandchildren [on that plane]?" (Monroe, 2017).

The *Intergovernmental Panel on Climate Change* 2022 report warned that increasing climate extremes are already pushing natural and human systems beyond their capacity to adapt (Pörtner et al., 2022). The *Guardian* called it "the bleakest warning yet," emphasizing that the window to secure a livable future is rapidly closing (Harvey, 2022). Research led by Johan Rockström has identified nine planetary boundaries that are crucial to the stability of the Earth. Six of them have already been crossed: climate change, biodiversity loss, land system change, freshwater use, chemical pollution, and biogeochemical flows. Many scientists warn that this trajectory puts us on course for a sixth mass extinction (Richardson et al., 2023).

According to Bill McGuire, an emeritus professor of geophysics and the author of *Hothouse Earth*, we may have already passed the point of no return in terms of climate change. The conclusion of his analysis is gloomy: a child born in 2020 will inherit a far more volatile and unforgiving planet than their grandparents did (McKie, 2022).

While climate change casts a dark shadow over our future, the risk of nuclear war is another major reason why the *Doomsday Clock* is so close to midnight. In 2020, an article published in *Nature* described the possible catastrophic scenarios of a limited nuclear conflict. The paper explained that a nuclear war could inject massive amounts of soot and smoke into the upper atmosphere, blocking sunlight and drastically cooling the planet. This phenomenon is known as nuclear winter and could lower global temperatures for months or even years. The same article also discussed how the resulting drop in sunlight and temperature, along with disrupted rainfall patterns, could devastate global food production and lead to widespread famine (Witze, 2020).

Another article published in *Science* explained that a nuclear war could disrupt the global climate so severely that it would lead to the starvation of billions of people. While the exact effects remain uncertain, scientists say the results would almost certainly be devastating (Savitsky, 2022). An additional work published in *Nature Food* estimated that "more than 5 billion people could die as a result of a war between the United States and Russia" (Xia et al., 2022).

Some very privileged, ultra-wealthy people like Elon Musk have entertained the idea of developing a "backup" plan for civilization when planet Earth will become uninhabitable. The plan involves the development of *SpaceX* Starship rockets to carry people to Mars. However, these are no more than delusional fantasies.

As the *Federation of American Scientists* explained, despite some progress in reducing nuclear arsenals since the Cold War, the world's combined stockpile of nuclear warheads remains at a very worrisome level. About 90% of all nuclear warheads are held by Russia and the United States (Kristensen et al., 2025). Both nations cast a dangerous shadow over the security of the world. Russian imperialism has clearly shown itself to be a threat to world peace, not only after the recent invasion of Ukraine, but also during previous wars in Afghanistan, Chechnya, and Georgia as well as past interventions in Eastern Europe.

American imperialism, however, is a far greater threat to global stability. As Jeffrey Sachs, professor at *Columbia University* and renowned expert in economics and world affairs, put it, NATO, which is largely dominated by the U.S., is "a clear and present danger to world peace, a war machine run amok" (Sachs, 2024). The U.S. accounts for 39% of global military spending with 800 military bases in over 70 countries. A comprehensive report on its military activities shows that the U.S. has been responsible for a historically unprecedented number of "military operations" around the world (Plagakis & Torreon, 2023).

It is also a source of political destabilization, such as organizing or facilitating coups, funding terrorism, and providing military support to brutal dictators such as Mobutu, Saddam Hussein, Suharto, and Pinochet.

Alongside the United States, Israel now stands as one of the greatest threats to world peace. This stems not only from the ongoing genocide and its powerful lobbying influence on U.S. policies, but also from its repeated aggression against other sovereign nations. In early September 2025, Israel attacked six countries within just 72 hours: Palestine, Lebanon, Syria, Tunisia, Yemen, and Qatar. Most dangerous of all, however, was Israel's strike on Iran on June 12–13, 2025, which pushed the world into even greater peril. Even now, the risk of escalation, with the potential involvement of Western powers and Iran's strategic allies, including Russia, and to some extent China, casts a long shadow over global security.

2.2   *No Environmentalism without Pacifism*

The climate crisis and the risk of nuclear war are inextricably interconnected. The climate crisis is a major risk factor for current and future wars. In turn, wars greatly exacerbate environmental degradation and render current global sustainability efforts virtually meaningless. As highlighted in an article published in the *Proceedings of the National Academy of Sciences*, higher temperature scenarios have the potential to create systemic risk and a cascade of impacts on different sectors of society that, at even modest levels of warming, can lead to what the authors define as "climate endgame" (Kemp et al., 2022).

Similarly, the *Global Risks Report 2023*, a publication of the *World Economic Forum*, warned of a potential "poly-crisis" related to climate impact on natural resources such as food and water, and the associated "socio-economic and environmental impacts." The report stated that there are "emerging or rapidly accelerating risks to natural ecosystems, human health, security ... and economic stability that could become crises and disasters in the next decade" (World Economic Forum, 2023).

A recent report by the *University of Exeter* and *Institute and Faculty of Actuaries* (IFoA) warns that the global economy could experience a 50% loss in GDP between 2070 and 2090 if immediate action is not taken to combat climate change. The potential collapse is attributed to severe climate impacts, including wildfires, floods, droughts, and rising temperatures. Without action, there could be significant socio-political fragmentation worldwide, failure of states with resulting rapid, enduring, and significant loss of capital, and unprecedented public health consequences (Laville, 2025).

A climate change-driven poly-crisis and its impacts on access to food and water can lead to trade wars, violence, and conflict between nations. There is already ample evidence that climate change is a major driver of conflict and political violence (Hendrix et al., 2023). These risks may escalate in the future as climate change forces nations to compete for dwindling natural resources (Dyer, 2010). When these crises generate national confrontation between nations endowed with nuclear weapons, the threat becomes truly existential.

If climate change creates wars, wars create more climate change.

The full extent of environmental damage in Gaza has not yet been fully assessed, but analyses of satellite imagery in March 2024 showed the destruction of some 38–48% of tree cover and agricultural land. The survey showed that olive groves and farms have been turned into barren land, with soil and groundwater contaminated by munitions and toxins. The sea has been overwhelmed with sewage and waste, while the air has become thick with smoke and particulate pollution. The relentless bombardment has had an enormous impact on Gaza's ecosystems and biodiversity, to the point that some scientists consider it a sign of "ecocide," a possible additional war crime (Ahmed et al., 2024).

A study entitled *A Multitemporal Snapshot of Greenhouse Gas Emissions from the Israel-Gaza Conflict* estimated that emissions from the first six months of military operations were greater than the annual emissions of 26 individual countries and territories (Otu-Larbi et al., 2024).

Russia's invasion of Ukraine too has generated devastating environmental consequences. An analysis of the conflict-related climate impacts of the Russia-Ukraine war found that it generated more emissions than 175 countries do in a year. Another study estimated the impact of the destruction of the *Nord Stream 1* and *Nord Stream 2* gas pipelines near the Danish island of Bornholm in the Baltic Sea in September 2022, for which German authorities issued an arrest warrant for a Ukrainian national (Aitken, 2025). The results showed that more than 115,000 tons of natural gas were released in the six days following the explosions, contributing to greenhouse gas emissions equivalent to one-third of Denmark's total annual CO_2 emissions (Sanderson et al., 2023). While wars have significant impacts on greenhouse gas emissions, they remain largely overlooked in international climate change agreements, a gap some authors have termed the "military blind spot" (Belcher et al., 2020).

The interrelationship between climate change and war demonstrates the limits of single-issue advocacy for environmental or pacifist causes and highlights the need for systemic change in all sectors of society.

Every serious environmentalist must also be a pacifist. There can be no peace without sustainability, and no sustainability without peace.

2.3 *War and Peace: Cooperation or Annihilation?*

Just as climate change is more than an environmental crisis, the risk of nuclear war is far more than a geopolitical problem. These two existential threats are symptoms of deep fractures in social relationships within and between countries and of systemic dysfunctions at the heart of our model of human development and culture.

Ultimately, the climate crisis and the risk of nuclear war are both crises of cooperation.

As the scientists working on the *Doomsday Clock* wrote: "Leaders around the world must immediately engage in renewed cooperation in the many ways and venues available to reduce existential risk. The world's citizens can and must organize to demand that their leaders do so and do so quickly" (Bulletin of the Atomic Scientists, 2025).

These lines echo those written more than seventy years ago by Bertrand Russell and Albert Einstein in their anti-nuclear war manifesto: "In the tragic situation facing humanity ... consider yourselves only as members of a biological species that has had a remarkable history and whose disappearance none of us wishes to see ... People barely realize that the danger is to themselves, their children and their grandchildren, and not to a vaguely frightened humanity ... we must remember that ... the question between East and West ... must not be decided by war ... This then is the question we ask you, rigid, terrifying, inevitable: will we put an end to the human race, or will humanity give up war?" (Russell & Einstein, 1955).

These words still ring true today.

At a time of escalating military spending and proxy confrontations between nuclear superpowers, we must seriously and profoundly reexamine our potential for mutual aid and reciprocity with distant and diverse peoples and countries. The threats of nuclear war and climate change require an unprecedented level of global partnership within and among nations. It also requires urgent political and economic transformations. However, some changes must also occur within us and our ability to cooperate and connect with strangers, distant people, foreigners and even enemies.

Although Darwinian vulgarizers would have us believe that the fittest are those who successfully compete in a state of war and confrontation, evolution tells us a different story. Species that are incapable of mutual understanding, cooperation, and peaceful coexistence risk self-destruction. This has never been truer for Homo sapiens than it is now.

As a sentence often misattributed to zoologist Alfred Emerson says: "The issue is clear: it is cooperation or annihilation."

3　　　Between Good and Evil

In 1963, the *Great Ape House* at the *Bronx Zoo* in New York featured an art installation consisting of a mirror placed beside the gorillas and orangutans. Beneath the mirror was an inscription that read: "You are looking at the most dangerous animal in the world. It alone of all the animals that ever lived can exterminate (and has) entire species of animals. Now it has the power to wipe out all life on earth."

This installation is a stark reminder that we, Homo sapiens, are far less peaceful than we would like to believe. History, biology, and evolution all suggest that our 300,000-year journey on Earth has been marked not only by extraordinary advances in science, technology, and civil rights, but also by war, conquest, and mass murder. Our collective power and unparalleled ability to adapt have come at a devastating cost: unceasing environmental destruction and the prolonged suffering of other animals. We have risen to dominance not through moral superiority, but through our capacity for organized violence and relentless ecological exploitation. In many ways, we have acted as the most systematic "serial ecological murderers" the planet has ever known.

Even in the modern era, despite unprecedented advances in health, quality of life, material standards and political organization, we continue to kill other species and each other on an astonishing scale. Technological and scientific breakthroughs of the past century have not limited this violence. They have amplified it. Our capacity for mass murder has grown alongside our scientific and technological progress.

Are we hopelessly hardwired for violence? Is aggression etched into our DNA?

3.1　　*Born To Be Selective*

Cooperation in the pursuit of world peace depends on cultivating an identity that transcends communities and nations, one rooted in solidarity unclouded by stereotypes and cultural bias. But our ability to cooperate across cultures depends on the moral strengths and weaknesses of human nature, a mystery that continues to challenge scientists, activists, politicians, religious leaders, volunteers, and everyday people alike.

Evolutionary biology offers a useful starting point to understand human nature. In his influential book *The Selfish Gene*, Richard Dawkins introduced the provocative idea that we are driven by genetic imperatives or selfish impulses that exist not to preserve us, but to ensure the replication of our DNA. According to Dawkins, we are merely vehicles, biological machines, serving the interests of our genes. Genes do not care about us, but they use us. We do not care about them, yet they are our masters. And we obey (Dawkins, 2016).

Although there is compelling evidence that we are largely guided by egotistical motives and that war is deeply rooted in human nature, science also reveals that these are not our only defining traits. Humans have unique predatory tendencies, but are also endowed with social, cooperative, and moral capacities. It is true that we are the only species known to commit mass murder, torture, and genocide. But we are also deeply social animals: from an early age, humans exhibit an innate need for love, affection, and meaningful connection.

Psychologists have long recognized the importance of emotional bonds, especially those formed in early childhood with a mother figure. These bonds play a vital role in healthy psychological development. Without them, growth is impaired. Studies have shown that children deprived of maternal affection often exhibit anxiety, anger, fear of abandonment and heightened aggression (Zhang & Bayly, 2024). Psychologist David Levy described this deep emotional void as "primary affect hunger", a condition marked by an inability to feel or express a full range of emotions. This emotional numbness is especially pronounced in some orphans, who may appear hollow, detached, and incapable of forming genuine emotional connections (McLaughlin et al., 2014).

However, these cherished needs for maternal love and affection, which are fundamental to our development, do not make us inherently virtuous. According to evolutionary biologists, the desire for love and connection can be viewed just as adaptive tools for survival and reproduction. The same is true for our need to socialize and be with others, whether that be a family, group, or community. Humans evolved as social creatures because living in groups offered concrete advantages, such as safety, shared resources, and greater reproductive success. Like wolves that hunt in packs or birds that defend their nests, humans benefit from "strength in numbers." Our social nature is closely tied to the laws of group selection. Although modern societies no longer punish solitude with death, the impulse to seek community remains hardwired.

Evolutionary biology also provides an explanation for our altruism and empathy. Most of us like to think of ourselves as compassionate creatures and believe that acts of generosity disprove the idea that we are merely machines guided by selfish genes. However, even our most generous and altruistic behaviors can be interpreted as evolutionary strategies designed to maximize chances of survival and reproduction (Dawkins, 2016).

Neuroscience has shown that acts of generosity, kindness and engagement in social causes activate the mesolimbic reward system in the brain. This system is also stimulated by food, sex, drugs, and money (Moll et al., 2006). This may be interpreted as corroboration that we are not inherently good and that we adopt generous acts because we feel emotionally rewarded when we are generous. In other words, we do good things not because we are good, but just because it makes us feel good.

This evolutionary lens may also help to explain the selective nature of our empathy. Research suggests that our capacity for compassion is shaped, and constrained, by genetic predispositions that favor in-group bias (Fu et al., 2012). In evolutionary terms, it makes sense to care more for a sibling than for a stranger, or to feel stronger loyalty to those who share our language, religion, or political beliefs. These parochial instincts are not cultural accidents, but biological inheritances, expressions of a deep-seated tribalism etched into our DNA.

The same instinct explains why our empathy weakens with emotional or physical distance. We are simply not wired to feel the same compassion for a distant stranger as we do for a family member or fellow citizen. From the perspective of natural selection, helping outsiders does not enhance genetic success; therefore, we rarely do so.

Curiously, although selective empathy seems to be an inevitable psychological trait, a result of our evolutionary heritage, most people are not particularly proud of it. Studies in social psychology show that most individuals value egalitarian empathy or the idea that compassion should extend to all people, regardless of identity or affiliation. People consistently rate universal empathy as more morally admirable than parochial or tribal emotion. Most of us view people who exhibit a broader type of compassion more positively, and view those driven by narrow, in-group favoritism with suspicion or disapproval (Fowler et al., 2021).

In other words, evolution may explain our instincts, but not our ideals and aspirations. While biology may have wired us for selective empathy, our moral imagination asks us to transcend it. In a sense, we are all upgraded primates, still governed by tribal instincts, yet aspiring to be more.

Paraphrasing a famous line in Oscar Wilde's play *Lady Windermere's Fan* (Wilde, 1892), we may all be "in the gutter", and yet we are "looking at the stars."

3.2   *Fair Instincts*

The Russian anarchist, geographer, and evolutionary theorist Peter Kropotkin, whom some have called the "Jesus Christ of Russia," was one of the first thinkers to systematically challenge the idea that humans are primarily selfish, competitive, and warlike. Although he deeply admired Charles Darwin, Kropotkin believed Darwin's theories pointed to a different conclusion: that cooperation, not conflict, is the foundation of human progress.

Born into an aristocratic family in Moscow in 1842, Prince Kropotkin was a rebellious and restless spirit. Despite repeated imprisonments for his revolutionary activities, he remained steadfast in his commitment to what may perhaps be called "universal love." His central thesis was simple but powerful: mutual help, not individual struggle, is the dominant force in evolution.

In his most influential book, *Mutual Aid*, Kropotkin argued that many animal societies, ants, bees, and others, survive not through competition but through reciprocity and collective support (Kropotkin, 2012). Contrary to Hobbes' view of life as a "war of all against all," Kropotkin observed that nature frequently rewards peaceful coexistence. He extended this insight to human evolution, arguing that cooperation, not only among kin but also across communities, has been crucial to our survival.

Kropotkin's work was inherently political. In stark contrast to Herbert Spencer's Social Darwinism, which glorified competition and justified inequality, Kropotkin envisioned a decentralized, stateless society rooted in voluntary cooperation and shared resources. In *Ethics: Origin and Development*, he traced moral evolution from early human societies to the modern age, arguing that morality arises naturally from social interaction, not imposed laws or religious edicts (Kropotkin, 2021).

Unlike some of his fellow revolutionaries, however, Kropotkin championed both social justice and individual freedom. He opposed not only capitalism but also authoritarian communism, believing that all forms of hierarchical control can lead to oppression. He believed in self-governance and a form of libertarian socialism grounded in nature itself. In his opinion, *if* we could put ourselves in the place of others, so that we could feel others as if they were ourselves, we would no longer need rules, laws. People would simply act for the common good, and so we would never do anything against another whom we would feel to be ourselves.

Though his vision may appear idealistic, contemporary science supports some of his claims. Modern evolutionary psychology and neuroscience research increasingly affirm that mutual aid is deeply rooted in our biology (Decety, 2015). As Oren Harman, author of *The Price of Altruism*, wrote, "Nature had made cooperation its chief executive" (Harman, 2010).

Research also shows that we have an innate sense of justice (Wang et al., 2019). Research on sense of fairness suggests that social justice is not just a dream of radical thinkers, but a trait embedded in our DNA. Studies across species and cultures confirm that fairness is a universal human value, transcending politics, religion, and geography.

Feelings of fairness are not only in our brains, but also in our guts. Part of our morality is, in fact, rooted in intuition. This both contradicts and complements Lawrence Kohlberg's theory of morality, which showed that our ethical qualities develop in stages and that cultivating a critical conscience requires reason and reflection. However, when faced with a moral dilemma, the amygdala, a brain structure central to processing emotions such as fear and anger, is usually activated before the dorsolateral prefrontal cortex, a region of the brain that plays a key role in cognitive control, decision-making, and rational thinking, especially in situations that require one to override emotional impulses (Bavel et al., 2012).

Research also shows that we are not alone in the animal world in our ability to perceive injustice. A famous paper published in *Nature*, titled *Monkeys reject unequal pay*, found that capuchin monkeys, highly social animals, react negatively to unfairness. They became uncooperative and showed signs of anger when exposed to unequal treatment in an experiment in which grapes and cucumbers were used as rewards for sharing a small stone or token (Brosnan & de Waal, 2003). Similar behaviors have been observed in pigs and birds. A 2023 study found that pigs react to the distress of their peers and seek comfort from others, suggesting emotional contagion and social support (Zhang et al., 2023). Ravens have also shown affiliative behavior. After conflicts, bystanders often comfort victims, especially those with whom they share close social bonds. This behavior suggests that ravens are sensitive to the emotions of others and may act to provide support, reflecting a form of empathic response (Fraser & Bugnyar, 2010).

Findings from developmental psychology further support the idea that a sense of fairness is innate. Research indicates that even young children possess a rudimentary sense of fairness early in life, which is a universal feature of human development (Surian et al., 2025). This is already evident in everyday interactions, for example, when a child spontaneously protests, "that's not fair!" upon perceiving unequal or unfavorable treatment. Cross-cultural research confirms this hypothesis. A study examined how children from seven different countries, ranging in age from four to fifteen, responded to scenarios involving unequal distribution of resources. The findings showed that aversion to inequity, or receiving less than a peer, was consistently observed across cultures and typically emerged around age four (McAuliffe et al., 2017).

Some critics, however, argue that these experiments do not demonstrate a truly evolved sense of fairness, but rather a self-interested response to unfair treatment. A true egalitarian sense of fairness requires recognizing injustice and trying to redress inequality when others are treated unfairly. A true sense of justice may involve paying a personal price to punish dishonesty and cooperating altruistically even when there is nothing to gain and everything to lose.

Several research studies have examined this particularly evolved type of sense of fairness. These studies have clearly shown that we feel a sense of indignation not only when we experience unfair treatment that harms us, but also when we witness acts of injustice against others. They also showed that most people are willing to punish others in order to restore justice or sanction unfair behavior even at a personal cost (Fehr & Gächter, 2002).

Even children are willing to penalize wrongdoers to restore justice. A study published in *Current Biology* found that children as young as three exhibit a natural inclination toward restorative justice, motivated by a deep concern for victims' well-being. In scenarios involving harm, these young children preferred actions that restored the victim's well-being rather than actions that solely punished the offender (Riedl et al., 2015). Another study confirmed that children evaluate restorative responses more positively than punitive ones. This preference for restorative justice over retribution seems to indicate not only children's inherent concern for the welfare of victims, but also a deep-seated desire to restore social harmony (Liu et al., 2021).

4   Imagine There Is No Other Land

> I don't believe in war
> I do not want to be buried under any flag.
> I want to be remembered for my dreams.
>
> VITTORIO ARRIGONI

Few things challenge the idea that humans are purely selfish beings driven only by survival and reproduction more profoundly than acts of self-sacrifice. While it is unsurprising that parents would risk their lives for their children, many of the most powerful examples of self-sacrifice go far beyond kinship or personal ties. They involve risking everything, not just for loved ones, but also for strangers, principles, or causes that offer no tangible reward.

What compels Western doctors to abandon the safety and privilege of a well-paid career at a comfortable medical center to work at a hospital under relentless bombardment, where every moment could be their last? What motivates university students to occupy campuses in solidarity with Gaza, risking expulsion and arrest, and jeopardizing their entire future? Why do some Israeli teenagers refuse military service, choosing prison over participating in a war they see as unjust? What drives some journalists to knowingly put themselves in mortal danger to report the truth from conflict zones like Gaza? What motivates professors to speak out for Palestine, even when it means suspension, termination, or permanent damage to their career? And what, ultimately explains Aaron Bushnell's decision to set himself on fire in front of the Israeli Embassy in Washington, D.C.? Is it biology alone?

Some evolutionary researchers might offer quick explanations for these seemingly paradoxical acts. After all, many animal species exhibit self-sacrificing behavior for the sake of group survival. But even then, perhaps deep down, we sense that these acts of heroism and solidarity resist purely rational or biological interpretation. Attributing them solely to the activation of a brain region or a genetic drive, or reducing them to evolutionary impulses or survival strategies, strips them of their true significance.

There is something profound, even sacred, about a selfless devotion to a distant, even abstract, social cause that no scientific model, evolutionary theory, or psychological study can fully contain or explain. It is a moral force from within, unseen, intangible, that compels some individuals to risk everything, to endure suffering, and at times, to lose their lives. Not for personal gain, but for justice, for the dignity of others, for a world they believe in, one they may never live to see.

While we all have a natural tendency to build barriers, to feel empathy selectively, and to privatize our love for those within our immediate circles, many people defy that instinct. They feel genuine outrage at violations of fairness and dignity involving complete strangers, distant others. It is no coincidence that fictional heroes like Robin Hood and Zorro, who defy the rich and powerful to protect the vulnerable, remain deeply embedded in the popular imagination. They resonate with our innate sense of justice and egalitarian empathy.

This does not necessarily suggest that human beings are naturally kind and peaceful, corrupted only by civilization or capitalism. Yet, we are not inherently selfish and violent either. While there is no shortage of ambitious theories about human nature, most fail to recognize the extent to which human beings are malleable. Selfish or altruistic genes do not determine who we are. As Gabor Maté put it: "The nature of our nature is not particularly constrained by our nature. Who we really are is and always will be an open question" (Maté, 2022, p. 17).

We are not purely altruistic or entirely self-serving, nor are we innately hierarchical or wholly egalitarian. We can be any of these things, sometimes simultaneously, depending on the circumstances. And crucially, those circumstances are shaped by our collective choices. While genetic research offers valuable insights into human behavior, its explanatory power is often overstated when weighed against environmental influences. We are shaped not only by our biological inheritance but also by education, socialization, and culture. Epigenetic mechanisms illustrate how genes are "turned on and off" in response to environmental circumstances. Factors such as socioeconomic conditions, stress, and the environment can all influence gene expression (Gillman et al., 2024). Even experiences of discrimination and structural racism can leave biological marks (Krieger et al., 2023).

There is nothing inherently selfish, aggressive, competitive, or warlike about human nature. These traits are frequently cultivated by the cultural, and environmental contexts we inhabit. While genes may shape our tendencies and predispositions, they do not dictate our destiny. We do. Our genetic blueprint is not a fixed script. It is fluid, constantly interacting with the world we construct through our choices and actions.

4.1 *We Are the World*

If the context shapes who we are, and we in turn shape the context, a fundamental question arises: can empathy transcend the boundaries of culture, nation, and tribe? Is there such a thing as universal empathy, distinct from the selective, parochial compassion we often reserve for those we perceive as part of our in-group?

Although relatively few studies have directly explored our capacity to empathize with distant and unfamiliar populations, the literature recognizes broader constructs such as *global empathy*, (Bachen et al., 2012) *ethnocultural empathy*, (Kapıkıran, 2023) and even *universal love* (Trent et al., 2020). These concepts suggest that the human capacity for empathy is not inherently limited to familiar or culturally similar others.

However, the scope of our empathy is often shaped by our social identity. Most people define themselves in relation to a group, be it family, community, nation, ethnicity, or even a sports team. From both evolutionary and psychological perspectives, this is adaptive: group affiliation fosters belonging, reinforces shared norms, and provides meaning through connection to something greater than the self.

Yet, the very mechanisms that unite us can also divide us. Group identity, especially when rooted in nationalism or rigid ideology, often works by emphasizing separation. The idea of "us" gains meaning in contrast to "them." This mindset can easily lead to dehumanization, diminishing our concern for those outside our group. The stronger our in-group attachment, the more likely we are to view outsiders as less deserving of dignity, compassion, or even basic humanity.

The notion of nationality, which only gained prominence with the advent of the modern nation-state, is now taken for granted. Yet it is a relatively recent development in the long history of Homo sapiens. Human civilization itself accounts for less than one percent of our species' 300,000-year existence; nationalism, even less. Former British Prime Minister Theresa May once claimed that "if you think you are a citizen of the world, you are a citizen of nowhere" (BBC News, 2016). One might argue the opposite: those who are citizens of nowhere may, in fact, be citizens of everywhere.

While most people tend to identify with their own communities, a minority cultivates a broader sense of social identity, one that sees all human beings as equally deserving of empathy and dignity. Two key concepts help explain this expanded moral perspective: cosmopolitanism and egalitarianism. Cosmopolitanism is "the idea that all humans can or should see themselves as belonging to a universal group that includes all people and should treat all people as having equal (rights and) moral standing" (Faulkner, 2018). Egalitarianism is "the belief system emphasizing that all individuals should have equal opportunities and treatment. It's a core tenet of many social and political movements, advocating for a society where everyone is treated with the same respect and afforded the same rights and opportunities" (Bidadanure & Axelsen, 2025).

Political psychologist Sam McFarland, author of *Heroes of Human Rights: Stories of Women and Men that Created Human Rights*, (McFarland, 2021) has contributed significantly to this field with the concept of *Identification with All Humanity* (Feng et al., 2022). This construct measures the degree to which individuals view themselves not only as part of a local or national community, but also as members of the global human family. Building on social identity theory, it highlights the role of empathy, compassion, and moral concern that extend beyond borders. When we identify with all of humanity, there are no longer "other" countries, because the only country is the world. There are no longer "foreigners," because all of humanity becomes our nation.

This is not merely an abstract, utopian idea: it is a measurable trait (Feng et al., 2022; Reysen & Hackett, 2016). McFarland developed an instrument that measures identification at three levels: community, nation, and all humanity. This construct has been widely used in studies that show correlations between global human identification and various outcomes including moral judgment, (Kahane et al., 2015) charitable behavior, (Law et al., 2022) and advocacy for human rights (Bassett & Cleveland, 2019). High identification with all humanity is also associated with morality and openness, empathy, secure attachment styles, (Hackett & Hamer, 2023) and an inclusive approach to social relationships such as welcoming people of different ethnic backgrounds into one's social circle (Hamer et al., 2021).

Conversely, a strong global identification is negatively associated with dehumanization, prejudice, and utilitarian callousness. It is also linked to lower levels of anxiety and depression, suggesting that seeing oneself as part of a shared humanity is not only ethically commendable, but also psychologically beneficial (McFarland et al., 2012).

The significance of these findings cannot be overstated. In an era defined by impending existential threats, polarization, tensions between nuclear super-powers, and national populism, an identity that transcends flags, borders, and tribal affiliations offers a glimpse into a more expansive moral imagination, and the potential for a collective psychological shift toward a deeper, broader humanitarianism.

As science fiction writer Isaac Asimov observed: "I don't believe in nations ... because Zionism merely sets up one more nation to trouble the world. It sets up one more nation to have "rights" and "demands" and "national security" and to feel it must guard itself against its neighbors." He continued: "There are no nations! There is only humanity. And if we don't come to understand that right soon, there will be no nations, because there will be no humanity" (Asimov, n.d.).

4.2   *Just Epiphanies*

"Inshallah Shalom", Peace if God Wills, is the title of an album by Yair Dalal, a former Israeli soldier turned musician. The phrase, an evocative fusion of Arabic (*Inshallah*—إن شاء الله) and Hebrew (*Shalom*—שָׁלוֹם) symbolizes a vision of coexistence and reconciliation. It is a deliberate gesture, both symbolic and sonic, bridging divides so often portrayed as irreconcilable.

I watched Yair perform and speak at a seminar in Venice. Between musical compositions on his oud, he shared the personal story that transformed his life: a moment he described as a turning point. It happened in Lebanon, during an Israeli military operation. Surrounded by militarism and destruction, he reached an inner reckoning and, unable to reconcile his actions with his conscience, he made a vow: "I am out of here. I'm going to become a music teacher."

Yair's story is revealing, but not unique. Across the Israeli–Palestinian divide, many individuals have chosen to reject hatred and violence, despite living under social systems that thrive on fear and dehumanization. Their acts echo moments like the Christmas Truce of World War I, when enemy soldiers met in peace across the trenches of the Western Front. Or when Italian and Austro-Hungarian soldiers in the Alps saluted instead of firing.

These ruptures in the machinery of war offer a glimmer of hope, proof of what is possible when propaganda, patriotism and nationalism no longer rule our hearts. Often, these moments arise suddenly, seemingly out of nowhere. They spring from a newfound clarity, a quiet revolt against narratives that divide and destroy. In such moments, individuals disobey out of awakening, an eruption of conscience, a break with the logic of hatred, a recognition of the humanity in those once seen as enemies.

These moments may be called epiphanies or epiphanies of justice. Or, more simply, just epiphanies.

Epiphany comes from the Greek ἐπιφάνεια (epiphaneia), meaning *appearance* or *manifestation*, a showing forth. In Christian tradition, the term marks the revelation of the divine to the Magi. But more broadly, epiphany signifies a moment of heightened awareness, when the ordinary is replaced by the extraordinary.

Epiphanies are experiences that shatter routine thinking and awaken new awareness. They can be triggered by crisis, loss, beauty, art, illness, or the stillness that follows despair. They leave lasting marks of transformation. They offer new, lateral angles of interpretation. Those who experience them often feel reborn, as if they have gained a new way of seeing themselves and the world.

While theologians, poets, and philosophers have long explored epiphanies, their power extends far beyond spiritual and mystical experience. Epiphanies are intensely personal and relevant to everyday life. Philosopher Sophie Grace Chappell writes that epiphanies "come from outside" and take us "out of ourselves" (Chappell, 2022).

We typically think of epiphanies as private experiences that transform individuals, not societies. And yet, in the face of the cruelty and indifference that have led to genocide and the existential threats now looming over us, perhaps what humanity needs most is a shared epiphany capable of awakening our collective conscience.

But is that realistic?

Perhaps not. But an even greater illusion is the belief that we can continue on the path we are on. Without a profound moral awakening, without the willingness to see the world through the eyes of others, our future hangs in the balance. Perhaps we do not fully realize what we are facing. Perhaps we are not entirely aware of what is happening. This is not about making a few adjustments here and there, but about taking an epochal leap in the way we perceive others.

We have grown accustomed to seeing enemies everywhere: in each other, in the pandemic, in climate change.

But perhaps the real enemy is not out there.

It is within us.

Like artists, we might learn to dream the impossible, to imagine what does not yet exist. We might begin to envision a new world with greater urgency than those in power cling to the existing, self-destructive status quo.

What comes next depends on the decisions we make now. We can choose peace or war, harmony or ruin. Never before has our survival teetered on a choice so vital or lethal.

References

Abbas, S., & Mitchell, L. (2024). Australian medical leadership's silence on Gaza is a moral failure. *The Lancet, 403*(10432), 1138–1139. https://doi.org/10.1016/S0140 -6736(24)00401-X.

ABC News. (2024, May 29). *Video Inside the moments leading up to death of 5-year-old in Gaza.* https://abcnews.go.com/International/video/inside-moments-leading -death-5-year-gaza-110628021.

Abdul, G. (2024, April 7). 'They're targeting healthcare workers': Airstrikes a constant fear for UK doctors in Gaza. *The Guardian.* https://www.theguardian.com/world/2024 /apr/07/targeting-healthcare-workers-airstrikes-constant-fear-uk-doctors-gaza.

Abu Alouf, R., & Slow, O. (2023, October 9). *Gaza 'soon without fuel, medicine and food'—Israel authorities.* BBC News. https://www.bbc.co.uk/news/world-middle -east-67051292.

Abuelaish, I. (2011). *I Shall Not Hate: A Gaza Doctor's Journey.* Random House of Canada.

Agerholm, H. (2017, January 2). *Israeli officials 'backing shoot-to-kill policy against Palestinians'.* The Independent. https://www.independent.co.uk/news/world/mid dle-east/israel-shoot-to-kill-policy-palestinian-suspects-human-rights-watch-idf -soldiers-west-bank-gaza-strip-a7505486.html.

Ahmed, K., Gayle, D., & Mousa, A. (2024, March 29). *'Ecocide in Gaza': Does scale of environmental destruction amount to a war crime?* The Guardian. https://amp.the guardian.com/environment/2024/mar/29/gaza-israel-palestinian-war-ecocide -environmental-destruction-pollution-rome-statute-war-crimes-aoe?fbclid=Iw ZXhobgNhZWoCMTAAAR0Mp7Xx2ddAHgotnOb_EuDw53voMz67ViLzA6yO p4jMj5OUjdAkWITWAEw_aem_ASUryOJg-Ksh9Kf6aVGoNxTxooX7dI92Zf7 _OCjyovnrQU1onRMt9mT9dHD5kBxQxDFreQmS7cgrv7HSWc65cpbH.

Ahmed, W. (2025, January 27). *Leaked Data Reveals Massive Israeli Campaign to Remove Pro-Palestine Posts on Facebook and Instagram.* https://www.dropsitenews.com/p /leaked-data-israeli-censorship-meta.

Airwars. (2023, October). *Airwars Gaza Patterns of Harm.* https://gaza-patterns-harm .airwars.org/.

Aitken, P. (2025, August 15). *Germany issues arrest warrant for Ukrainian suspect in connection to Nord Stream pipeline explosion: Reports.* New York Post. https://nypost .com/2024/08/15/world-news/germany-issues-arrest-warrant-for-ukrainian-sus pect-in-connection-to-nord-stream-pipeline-explosion-reports.

Ajana, B., Connell, H., & Liddle, T. (2024). *"It Could Have Been Us": Media frames and the coverage of Ukrainian, Afghan and Syrian refugee crises | SN Social Sciences. 4,* 135.

Akkerman, N. N. B., Mark. (2024, December 13). *Partners in Crime | Transnational Institute.* https://www.tni.org/en/publication/partners-in-crime-EU-complicity -Israel-genocide-Gaza.

Al Jazeera. (2024a, January 12). *Which countries back South Africa's genocide case against Israel at ICJ?* Al Jazeera. https://www.aljazeera.com/news/2024/1/9/which -countries-back-south-africas-genocide-case-against-israel-at-icj.

Al Jazeera. (2024b, June 21). *'The Night Won't End': Biden's War on Gaza | Gaza | Al Jazeera.* https://www.aljazeera.com/program/fault-lines/2024/6/21/the-night -wont-end-bidens-war-on-gaza-2.

Al Jazeera. (2024c, December 12). *UNGA demands permanent ceasefire in Gaza: How did your country vote? | Israel-Palestine conflict News.* https://www.aljazeera.com /news/2024/12/12/un-demands-permanent-ceasefire-in-gaza-how-did-your-coun try-vote.

Al Jazeera. (2025a, January 28). *Five-year-old among two killed in Israeli attacks amid Gaza ceasefire.* Al Jazeera. https://www.aljazeera.com/news/2025/1/28/five -year-old-among-two-killed-in-israeli-attacks-amid-gaza-ceasefire.

Al Jazeera. (2025b, April 3). *Gaza faces 'largest orphan crisis' in modern history, report says.* Al Jazeera. https://www.aljazeera.com/news/2025/4/3/gaza-faces -largest-orphan-crisis-in-modern-history-report-says.

Ali, L. (2022, March 2). *In Ukraine reporting, Western press reveals grim bias toward 'people like us'.* Los Angeles Times. https://www.latimes.com/entertainment-arts /tv/story/2022-03-02/ukraine-russia-war-racism-media-middle-east.

Allam, L., & Evershed, N. (2025, February 23). More than 10,000 First Nations people killed in Australia's frontier wars, final massacre map shows. *The Guardian.* https://www.theguardian.com/australia-news/2025/feb/23/more-than-10000-first -nations-people-killed-in-australias-frontier-wars-final-massacre-map-shows -ntwnfb.

Amnesty International. (2022, February 1). *Israel's apartheid against Palestinians.* Amnesty International. https://www.amnesty.org/en/latest/campaigns/2022/02 /israels-system-of-apartheid/.

Amnesty International. (2024, December 5). *Amnesty International concludes Israel is committing genocide against Palestinians in Gaza.* https://www.amnesty.org/en /latest/news/2024/12/amnesty-international-concludes-israel-is-committing-geno cide-against-palestinians-in-gaza/.

ANSA. (2024, January 11). *Israel has hit civilians in Gaza but no genocide—Tajani— Politics—Ansa.it.* Agenzia ANSA. https://www.ansa.it/english/news/politics/2024 /01/11/israel-has-hit-civilians-in-gaza-but-no-genocide-tajani_05247bb7-3385-42c9 -a907-565bdba33c12.html.

Applebaum, A. (2002, January 12). *Kill the Messenger Why Palestine radio and TV studios are fair targets in the Palestine/Israeli war.* Slate. https://slate.com/news-and -politics/2002/01/targeting-radio-and-tv-stations.html.

Arendt, H. (1973). *The Origins of Totalitarianism.* Houghton Mifflin Harcourt.

Armstrong, M. (2019, June 17). *Infographic: The Wars With the Highest Annual Death Tolls*. Statista Daily Data. https://www.statista.com/chart/18375/the-wars-with-the-highest-annual-death-tolls.

Arnaout, A. (2024, January 5). *Another Israeli minister calls for encouraging Palestinians to leave Gaza.* https://www.aa.com.tr/en/middle-east/another-israeli-minister-calls-for-encouraging-palestinians-to-leave-gaza/3101423.

Asch, S. E. (1951). Effects of Group Pressure on the Modification and Distortion of Judgments. In *In Guetzknow, H., Ed., Groups, Leadership and Men* (pp. 177–190). Carnegie Press.

Asimov, I. (n.d.). *Quote by Isaac Asimov: "The Earth should not be cut up into hundreds of ..."* Retrieved 22 April 2025, from https://www.goodreads.com/quotes/93498-the-earth-should-not-be-cut-up-into-hundreds-of.

Athar, M. E. (2023). The Syndrome of Collective Callous-Unemotional Traits: Reflections from large group dynamics. *International Journal of Applied Psychoanalytic Studies, 20*(1), 83–97. https://doi.org/10.1002/aps.1774.

Australian Medical Association. (2024, April 24). *AMA statement on the conflict in Israel and Gaza.* Australian Medical Association. https://www.ama.com.au/articles/ama-statement-conflict-israel-and-gaza.

Ayoub, H. H., Chemaitelly, H., & Abu-Raddad, L. J. (2024). Comparative analysis and evolution of civilian versus combatant mortality ratios in Israel-Gaza conflicts, 2008–2023. *Frontiers in Public Health, 12.* https://doi.org/10.3389/fpubh.2024.1359189.

Bachen, C. M., Hernández-Ramos, P. F., & Raphael, C. (2012). Simulating REAL LIVES: Promoting Global Empathy and Interest in Learning Through Simulation Games. *Simulation & Gaming, 43*(4), 437–460. https://doi.org/10.1177/1046878111432108.

Bandura, A. (1999). Moral disengagement in the perpetration of inhumanities. *Personality and Social Psychology Review, 3*(3), 193–209. https://doi.org/10.1207/s15327957pspr0303_3.

Baron-Cohen, S. (2012). *The Science of Evil: On Empathy and the Origins of Cruelty.* Basic Books.

Bartov, O. (2024, August 13). As a former IDF soldier and historian of genocide, I was deeply disturbed by my recent visit to Israel. *The Guardian.* https://www.theguardian.com/world/article/2024/aug/13/israel-gaza-historian-omer-bartov.

Basile, M. (2024, May 12). *Gaza, l'Onu rivede al ribasso il numero delle vittime fra donne e bambini.* la Repubblica. https://www.repubblica.it/esteri/2024/05/12/news/gaza_lonu_rivede_al_ribasso_il_numero_delle_vittime_fra_donne_e_bambini-422919774/.

Bassett, J. F., & Cleveland, A. J. (2019). Identification With All Humanity, Support for Refugees and for Extreme Counter-Terrorism Measures. *Journal of Social and Political Psychology, 7*(1), Article 1. https://doi.org/10.5964/jspp.v7i1.678.

Baum, S. K. (2008). *The psychology of genocide: Perpetrators, bystanders, and rescuers* (pp. xi, 255). Cambridge University Press. https://doi.org/10.1017/CBO9780511819278.

Bavel, J. J. V., Packer, D. J., Haas, I. J., & Cunningham, W. A. (2012). The Importance of Moral Construal: Moral versus Non-Moral Construal Elicits Faster, More Extreme, Universal Evaluations of the Same Actions. *PLOS ONE*, 7(11), e48693. https://doi.org/10.1371/journal.pone.0048693.

Bayoumi, M. (2022, March 2). They are 'civilised' and 'look like us': The racist coverage of Ukraine. *The Guardian*. https://www.theguardian.com/commentisfree/2022/mar/02/civilised-european-look-like-us-racist-coverage-ukraine.

BBC. (1989, November 1). *Fascist Legacy*. https://www.imdb.com/it/title/tt1597615/.

BBC. (1993, July 21). *Lecture 5 titled "Speaking Truth to Power. The Reith Lectures, Edward Said—Representation of the Intellectual"*. BBC Radio 4. https://www.bbc.co.uk/programmes/p00gmx4c/episodes/player.

BBC. (2023a, July 20). *What is a war crime and could Putin be prosecuted over Ukraine?* https://www.bbc.com/news/world-60690688.

BBC. (2023b, November 6). Israel Gaza live news: UN says Gaza becoming a 'graveyard for children', as Israeli strikes intensify. *BBC News*. https://www.bbc.com/news/live/world-middle-east-67324897.

BBC. (2024, November 13). *Gaza surgeon describes drones targeting children*. BBC News. https://www.bbc.com/news/articles/c7893vpy2gqo.

BBC News. (2016, October 29). 'Mrs May, we are all citizens of the world,' says philosopher. *BBC News*. https://www.bbc.com/news/uk-politics-37788717.

BBC News. (2024, March 2). Aaron Bushnell: Friends struggle to comprehend US airman's Gaza protest death. *BBC News*. https://www.bbc.com/news/world-us-canada-68455401.

Beauchamp, Z. (2014, July 31). *95 percent of Jewish Israelis support the Gaza war | Vox*. https://www.vox.com/2014/7/31/5955077/israeli-support-for-the-gaza-war-is-basically-unanimous.

Beckett, L. (2024, May 10). Nearly all Gaza campus protests in the US have been peaceful, study finds. *The Guardian*. https://www.theguardian.com/us-news/article/2024/may/10/peaceful-pro-palestinian-campus-protests.

Belcher, O., Bigger, P., Neimark, B., & Kennelly, C. (2020). Hidden carbon costs of the "everywhere war": Logistics, geopolitical ecology, and the carbon boot-print of the US military. *Transactions of the Institute of British Geographers*, 45(1), 65–80. https://doi.org/10.1111/tran.12319.

Berg, M. (2024, May 6). *'You have been warned': GOP senators caution ICC over Israeli arrest warrants*. Politico. https://www.politico.com/live-updates/2024/05/06/congress/gop-senators-warn-icc-on-israel-00156232.

Berg, S. R., & Moench, M. (2025, February 8). *Hamas frees three Israeli hostages as Palestinian prisoners released*. https://www.bbc.com/news/articles/c4g9vyz747eo.

Bergman, R., & Kingsley, P. (2024, January 29). Details Emerge on U.N. Workers Accused of Aiding Hamas Raid. *The New York Times*. https://www.nytimes.com/2024/01/28/world/middleeast/gaza-unrwa-hamas-israel.html.

Berndsen, M., & Tiggemann, M. (2020). Multiple versus single immoral acts: An immoral person evokes more schadenfreude than an immoral action. *Motivation and Emotion, 44*(5), 738–754. https://doi.org/10.1007/s11031-020-09843-5.

Bhat, S. (2024, February 21). *'Proud of ruins of Gaza': Israeli minister rejoices at Palestine's distress*. 'Proud of Ruins of Gaza': Israeli Minister Rejoices at Palestine's Distress. https://www.trtworld.com/middle-east/proud-of-ruins-of-gaza-israeli-minister-rejoices-at-palestines-distress-17079853.

Bhutta, Z. A., Dominguez, G. B., & Wise, P. H. (2024). When is enough, enough? Humanitarian rights and protection for children in conflict settings must be revisited. *BMJ, 386*, e081515. https://doi.org/10.1136/bmj-2024-081515.

Bidadanure, J., & Axelsen, D. (2025). Egalitarianism. In E. N. Zalta & U. Nodelman (Eds.), *The Stanford Encyclopedia of Philosophy* (Spring 2025). Metaphysics Research Lab, Stanford University. https://plato.stanford.edu/archives/spr2025/entries/egalitarianism/.

Bigg, M. M., & Gupta, G. (2024, January 25). What to Know About the U.N. Court's Initial Ruling in the Genocide Case Against Israel. *The New York Times*. https://www.nytimes.com/2024/01/25/world/middleeast/icj-israel-genocide-ruling.html.

Blair, R. J. R. (2005). Responding to the emotions of others: Dissociating forms of empathy through the study of typical and psychiatric populations. *Consciousness and Cognition, 14*(4), 698–718. https://doi.org/10.1016/j.concog.2005.06.004.

Bloom, P. (2017). *Against Empathy: The Case for Rational Compassion*. Random House.

Blumenthal, M., & Maté, M. B., Aaron. (2024, January 10). *Screams without proof: Questions for NYT about shoddy 'Hamas mass rape' report—The Grayzone*. https://thegrayzone.com/2024/01/10/questions-nyt-hamas-rape-report/, https://thegrayzone.com/2024/01/10/questions-nyt-hamas-rape-report/.

Boffey, D., Jones, S., Cousins, R., & Wishart, E. (2023, October 13). Israel's darkest day: The 24 hours of terror that shook the country. *The Guardian*. https://www.theguardian.com/world/2023/oct/13/israel-darkest-day-24-hours-of-terror-hamas-gaza.

Bologna, M., & Aquino, G. (2020). Deforestation and world population sustainability: A quantitative analysis. *Scientific Reports, 10*(1), Article 1. https://doi.org/10.1038/s41598-020-63657-6.

Borger, J. (2024, December 11). Death feels imminent for 96% of children in Gaza, study finds. *The Guardian*. https://www.theguardian.com/world/2024/dec/11/death-feels-imminent-for-96-of-children-in-gaza-study-finds.

Bowcott, O., & Jones, S. (2008, January 23). *Johnson's 'piccaninnies' apology | Politics | The Guardian*. https://www.theguardian.com/politics/2008/jan/23/london.race

Brambilla, M., & Riva, P. (2017). Predicting pleasure at others' misfortune: Morality trumps sociability and competence in driving deservingness and schadenfreude. *Motivation and Emotion, 41*(2), 243–253. https://doi.org/10.1007/s11031-016-9594-2.

Brawn, S. (2024, October 24). *UK involvement in Israeli military operations revealed in Al Jazeera report*. The National. https://www.thenational.scot/news/24675646 .al-jazeera-probe-shows-extent-uk-assistance-israel/.

Breithaupt, F. (2019). *The Dark Sides of Empathy*. Cornell University Press.

Brett, A. (2015). 'The miserable remnant of this ill-used people': Colonial genocide and the Moriori of New Zealand's Chatham Islands. *Journal of Genocide Research, 17*(2), 133–152. https://doi.org/10.1080/14623528.2015.1027073.

Brosnan, S. F., & de Waal, F. B. M. (2003). Monkeys reject unequal pay. *Nature, 425*(6955), 297–299. https://doi.org/10.1038/nature01963.

Brown, D. K., & The Conversation. (2024, May 7). *'Protest Paradigm' Shows What's Wrong with Media Coverage of Student Activism*. Scientific American. https:// www.scientificamerican.com/article/protest-paradigm-shows-whats-wrong-with -media-coverage-of-student-activism/.

Bruneau, E. G., Cikara, M., & Saxe, R. (2017). Parochial Empathy Predicts Reduced Altruism and the Endorsement of Passive Harm. *Social Psychological and Personality Science, 8*(8), 934–942. https://doi.org/10.1177/1948550617693064.

B'Tselem. (2024, August). *Welcome to Hell: The Israeli Prison System as a Network of Torture Camps*. B'Tselem. http://www.btselem.org/publications/202408_welcome _to_hell.

Bulletin of the Atomic Scientists. (2025, January 28). 2025 Doomsday Clock Statement. *Bulletin of the Atomic Scientists*. https://thebulletin.org/doomsday-clock/2025 -statement/.

Cameron, C. D., Conway, P., & Scheffer, J. A. (2022). Empathy regulation, prosociality, and moral judgment. *Current Opinion in Psychology, 44*, 188–195. https://doi .org/10.1016/j.copsyc.2021.09.011.

Cameron, C. D., Harris, L. T., & Payne, B. K. (2016). The emotional cost of humanity: Anticipated exhaustion motivates dehumanization of stigmatized targets. *Social Psychological and Personality Science, 7*(2), 105–112. https://doi.org/10.1177 /1948550615604453.

Camut, N., & Boonen, E. (2023, February 23). 'We are all Ukrainian.' How the yellow -and-blue flag won over Europe. *Politico*. https://www.politico.eu/article/ukraine -russia-war-vladimir-putin-volodymyr-zelenskyy-emmanuel-macron-yellow-and -blue-flag-won-over-europe/.

CBS Sunday Morning (Director). (2024, May 5). *Remembering the October 7 attacks and 'The Moment Music Stood Still'* [Video recording]. https://www.youtube.com /watch?v=9ougIIirRXo.

Centre For Media Monitoring, M. (2024, March 6). *Center for Media Monitoring report 'Media Bias: Gaza 2023–24'*. https://cfmm.org.uk/cfmm-report-media-bias-gaza -2023-24/.

Chang, K.-K., & Zeldes, G. A. (2006). Three of Four Newspapers Studied Favor Israeli Instead of Palestinian Sources. *Newspaper Research Journal, 27*(4), 84–90. https:// doi.org/10.1177/073953290602700407.

Chappell, S. G. (2022). Introducing Epiphanies. In S. G. Chappell (Ed.), *Epiphanies: An Ethics of Experience* (p. 0). Oxford University Press. https://doi.org/10.1093 /oso/9780192858016.003.0001.

Chomsky, N. (1986). *Knowledge of Language: Its nature, origin, and use*. Praeger.

Chomsky, N. (1993). *Year 501: The Conquest Continues*. Verso.

Chomsky, N. (2006, April 25). *On the NATO Bombing of Yugoslavia, Noam Chomsky interviewed by Danilo Mandic*. RTS Online. https://chomsky.info/20060425/.

Chu, M. (2016, July 28). *The 3 Stages Of Truth In Life | HuffPost Life*. The Huffington Post. https://www.huffpost.com/entry/the-3-stages-of-truth-in_b_11244204.

Cohen, M. (2024, December 19). *Can Genocide Studies Survive a Genocide in Gaza?* Jewish Currents. https://jewishcurrents.org/can-genocide-studies-survive-a-geno cide-in-gaza.

Cohen/Walla, M. (2024, January 3). *Israel-Hamas War: 1,600 IDF soldiers suffer from combat PTSD*. The Jerusalem Post | JPost.Com. https://www.jpost.com/israel -hamas-war/article-780584.

Committee to Protect Journalists. (2025, October 13). *Journalist casualties in the Israel-Gaza war—*. https://cpj.org/2025/02/journalist-casualties-in-the-israel-gaza-con flict/.

Confino, J. (2024, September 4). *'Psychopath' Netanyahu sacrificed my cousin for politi-cal gain, claims hostage relative*. Yahoo News. https://www.yahoo.com/news/psy chopath-netanyahu-sacrificed-cousin-political-171033188.html.

Conley, J. (2024, May 1). *Israeli Finance Minister Denounced for Calling for 'Total Annihilation' of Gaza | Common Dreams*. https://www.commondreams.org/news /smotrich-gaza-annihilation.

Conley, J. (2025a, March 24). *'Everything Is Being Crushed': Journalist Hossam Shabat's Last Story Before He Was Killed by Israel | Common Dreams*. https://www.common dreams.org/news/hossam-shabat.

Connolly, K. (2024, April 10). German university rescinds Jewish American's job offer over pro-Palestinian letter. *The Guardian*. https://www.theguardian.com/educa tion/2024/apr/10/nancy-fraser-cologne-university-germany-job-offer-palestine.

Conrad, J. (1988). *Heart of Darkness*. (Edited by Robert Kimbrough, 3rd ed., W. W. Norton&Company,).

Cook, J. (2024, March 28). *War on Gaza: We were lied into genocide. Al Jazeera has shown us how | Middle East Eye*. https://www.middleeasteye.net/big-story/gaza-war-lies-genocide-al-jazeera-shown-how.

Cook, J., Oreskes, N., Doran, P. T., Anderegg, W. R. L., Verheggen, B., Maibach, E. W., Carlton, J. S., Lewandowsky, S., Skuce, A. G., Green, S. A., Nuccitelli, D., Jacobs, P., Richardson, M., Winkler, B., Painting, R., & Rice, K. (2016). Consensus on consensus: A synthesis of consensus estimates on human-caused global warming. *Environmental Research Letters*, *11*(4), 048002. https://doi.org/10.1088/1748-9326/11/4/048002.

Cope, J. (2024, March 30). *The latest blood libels against Israel—Opinion*. The Jerusalem Post | JPost.Com. https://www.jpost.com/opinion/article-794255.

Crimston, D., Hornsey, M. J., Bain, P. G., & Bastian, B. (2018). Toward a psychology of moral expansiveness. *Current Directions in Psychological Science*, *27*(1), 14–19. https://doi.org/10.1177/0963721417730888.

Crossley, N. (2020). Consistency, Protection, Responsibility: Revisiting the Debate on Selective Humanitarianism. *Global Governance*, *26*(3), 473–499.

Dadouch, S., & Westfall, S. (2024, May 20). *What is the ICC, the global court Israel fears may indict Netanyahu?* Washington Post. https://www.washingtonpost.com/world/2024/04/29/icc-israel-warrants-gaza/.

Daily Mirror. (2023, October 20). *Israel, Western media normalise psychopathic behaviour—Opinion | Daily Mirror*. https://www.dailymirror.lk/opinion/Israel-Western-media-normalise-psychopathic-behaviour/172-269563.

Daily Sabah. (2014, July 14). *'Mothers of all Palestinians should also be killed,' says Israeli politician*. https://www.dailysabah.com/mideast/2014/07/14/mothers-of-all-palestinians-should-also-be-killed-says-israeli-politician.

Das, S. (2025). *Uncivilised: Ten Lies That Made the West*. Hodder & Stoughton.

Davies, S., Pettersson, T., Sollenberg, M., & Öberg, M. (2025). Organized violence 1989–2024, and the challenges of identifying civilian victims. *Journal of Peace Research*, *62*(4), 1223-1240. https://doi.org/10.1177/00223433251345636.

Davis, S. K., & Nichols, R. (2016). Does Emotional Intelligence have a "Dark" Side? A Review of the Literature. *Frontiers in Psychology*, *7*. https://doi.org/10.3389/fpsyg.2016.01316.

Dawkins, R. (2016). *The Selfish Gene*. Oxford University Press.

de Las Casas B. (1999). *Short Account of the Destruction of the Indies*. Penguin.

De Vogli R. La strage degli innocenti a Gaza: si può accusare Israele di genocidio? Fatto Quotidiano. October 20, 2023. https://www.ilfattoquotidiano.it/2023/10/20/a-gaza-la-strage-degli-innocenti-si-puo-accusare-israele-di-genocidio/7326688/.

De Vogli, R. (2023, October 20). *A Gaza la strage degli innocenti. Si può accusare Israele di genocidio?—Il Fatto Quotidiano*. https://www.ilfattoquotidiano.it/2023/10/20/a-gaza-la-strage-degli-innocenti-si-puo-accusare-israele-di-genocidio/7326688/.

De Vogli, R., & Birbeck, G. L. (2005). Potential Impact of Adjustment Policies on Vulnerability of Women and Children to HIV/AIDS in Sub-Saharan Africa. *Journal of Health, Population and Nutrition, 23*(2), 105–120.

De Vogli, R., & Ferretti, A. (2024, May 10). *Gaza e l'empatia selettiva del mondo occidentale.* Micromega. https://www.micromega.net/gaza-empatia-selettiva-del-mondo-occidentale/.

de Waal, F. B. M., & Preston, S. D. (2017). Mammalian empathy: Behavioural manifestations and neural basis. *Nature Reviews. Neuroscience, 18*(8), 498–509. https://doi.org/10.1038/nrn.2017.72.

De Vogli, R., Montomoli, J., Abu-Sittah, G., & Pappé, I. (2025). Break the selective silence on the genocide in Gaza. *The Lancet, 406*(10504), 688-689. https://doi.org/10.1016/S0140-6736(25)01541-7

Decety, J. (2015). The neural pathways, development and functions of empathy. *Current Opinion in Behavioral Sciences, 3*, 1–6. https://doi.org/10.1016/j.cobeha.2014.12.001.

Decker, F. (2024, May 1). *Joe Scarborough's Condescending, Irresponsible, and Ignorant Rant on Students Protesting Gaza War.* Common Dreams. https://www.commondreams.org/opinion/joe-scarborough-gaza-campus-protests.

Defense for Children Palestine. (2023, December 14). *Year-in-review: Israeli forces carry out genocide against Palestinian children.* Defense for Children Palestine. https://www.dci-palestine.org/year_in_review_israeli_forces_carry_out_genocide_against_palestinian_children.

Deliso, M. (2023, October 18). *What is Palestinian Islamic Jihad? Israel blames group for Gaza hospital blast.* ABC News. https://abcnews.go.com/International/palestinian-islamic-jihad-israel-blames-group-gaza-hospital/story?id=104076124.

Democracy Now! (2002, August 14). *Israel's First Lady of Human Rights: A Conversation with Shulamit Aloni.* Democracy Now! http://www.democracynow.org/2002/8/14/israels_first_lady_of_human_rights.

Democracy Now! (2003, September 24). *Scholar Norman Finkelstein Calls Professor Alan Dershowitz's New Book On Israel a "Hoax".* Democracy Now! http://www.democracynow.org/2003/9/24/scholar_norman_finkelstein_calls_professor_alan.

Democracy Now! (2023a). *"A Textbook Case of Genocide": Israeli Holocaust Scholar on Israel's Gaza Assault.* https://www.democracynow.org/2023/10/16/raz_segal_textbook_case_of_genocide.

Democracy Now! (2023b, November 13). *Hear One of Dr. Hammam Alloh's Last Interviews from Gaza Before His Death.* Democracy Now! https://www.democracynow.org/2023/11/13/remembering_hammam_alloh.

Democracy Now! (2024, June 5). *Meet the Palestinian Lawyer Censored by Columbia and Harvard.* Democracy Now! https://www.democracynow.org/2024/6/5/harvard_columbia_law_school_nakba_censorship.

Des Garennes, C. (2025, February 26). *Salaita prompted donors' fury*. The News-Gazette. https://www.news-gazette.com/news/salaita-prompted-donors-fury/article_7c90ef7b-f622-5492-9bab-ab7efeac1ce8.html.

Deutsch, A., & van den Berg, S. (2024, April 26). *Mass graves in Gaza: What do we know? | Reuters*. Reuters. https://www.reuters.com/world/middle-east/mass-graves-gaza-what-do-we-know-2024-04-25/.

Deutsch, J. (2020, June 8). *The Mass Psychopathy of Shamelessness: From Israel to the UN*. Socialist Project. https://socialistproject.ca/2020/06/the-mass-psychopathy-of-shamelessness-from-israel-to-the-un/.

Diamond, J. M. (1997). *Guns, Germs, and Steel: The Fates of Human Societies*. W. W. Norton.

Doctors Without Borders. (2024a, March 1). *People killed in Gaza while waiting for food |*. https://msf.org.au/article/statements-opinion/gaza-people-killed-while-waiting-food.

Doctors Without Borders. (2024b, April 1). *Gaza: "Complete devastation of the health care system"*. https://www.doctorswithoutborders.org/latest/gaza-complete-devastation-health-care-system.

Doctors Without Borders. (2024c, December 19). *MSF report exposes Israel's campaign of total destruction*. https://www.msf.org/msf-report-exposes-israel%E2%80%99s-campaign-total-destruction.

Dong, K. L. (2023). *SELECTIVE EMPATHY: A Guide to Overcoming Narcissistic Traits and Becoming More Empathetic*. Independently published.

Dunbar-Ortiz, R. (2023). *An Indigenous Peoples' History of the United States*. Beacon Press.

Durgham, N. (2025, April 8). *Israel's Smotrich says 'not even a grain of wheat' will enter Gaza*. Middle East Eye. https://www.middleeasteye.net/news/smotrich-says-not-even-grain-wheat-will-enter-gaza.

Dyer, G. (2010). *Climate wars: The fight for survival as the world overheats*. Oneworld Publications.

Ebrahim, N. (2024, October 21). *'He got out of Gaza, but Gaza did not get out of him': Israeli soldiers returning from war struggle with trauma and suicide*. CNN. https://www.cnn.com/2024/10/21/middleeast/gaza-war-israeli-soldiers-ptsd-suicide-intl/index.html.

Eger, E. E., & Weigand, E. S. (2017). *The Choice: Embrace the Possible*. Simon and Schuster.

Eglash, R. M. (2024, May 13). *UN revises Gaza death toll, almost 50% less women and children killed than previously reported | Fox News*. https://www.foxnews.com/world/un-revises-gaza-death-toll-50-less-women-children-killed-previously-reported.amp.

Eiland, G. (2023, October 12). *It's time to rip off the Hamas band-aid*. https://www.ynetnews.com/article/sju3uabba.

Einstein, A., & Arendt, H. (1948). *Letter to The New York Times. New Palestine Party. Visit of Menachen Begin and Aims of Political Movement Discussed.* http://archive.org/details/AlbertEinsteinLetterToTheNewYorkTimes.December41948.

Ekman, P., & Ekman, E. (2017). Is Global Compassion Achievable? In E. M. Seppälä, E. Simon-Thomas, S. L. Brown, M. C. Worline, C. D. Cameron, & J. R. Doty (Eds.), *The Oxford Handbook of Compassion Science* (p. 0). Oxford University Press. https://doi.org/10.1093/oxfordhb/9780190464684.013.4.

El-Affendi, A. (2024). The Futility of Genocide Studies After Gaza. *Journal of Genocide Research*, 0(0), 1–7.

Elizur, Y. (2024, December 23). *'When You Leave Israel and Enter Gaza, You Are God': Inside the Minds. ...* https://www.haaretz.com/2024-12-23/ty-article/.premium/when-you-enter-gaza-you-are-god-inside-the-minds-of-idf-soldiers-who-commit-war-crimes/00000193-f043-d354-a59f-ff670ac80000.

Ellison, S., & Andrews, T. M. (2022, February 28). 'They seem so like us': In depicting Ukraine's plight, some in media use offensive comparisons. *The Washington Post.* https://www.washingtonpost.com/media/2022/02/27/media-ukraine-offensive-comparisons/.

Epstein G. (2024, March 26). *Gaza Fatality Data Has Become Completely Unreliable | The Washington Institute.* https://www.washingtoninstitute.org/policy-analysis/gaza-fatality-data-has-become-completely-unreliable.

Eric Fromm. (1958, May 19). The Problem of Power in Israel. *Jewish Newsletter,.*

Espín, A. M., Correa, M., & Ruiz-Villaverde, A. (2022). Economics students: Self-selected in preferences and indoctrinated in beliefs. *International Review of Economics Education*, 39, 100231. https://doi.org/10.1016/j.iree.2021.100231.

Ettingermentum. (2024, October 25). *Israel's Government of Psychopaths.* https://www.ettingermentum.news/p/israels-government-of-psychopaths.

European Union. (2024, March 18). *Foreign Affairs Council: Press remarks by High Representative Josep Borrell upon arrival | EEAS.* https://www.eeas.europa.eu/eeas/foreign-affairs-council-press-remarks-high-representative-josep-borrell-upon-arrival-16_en.

Evangelista, C. (2024, May 22). *Liliana Segre: «Dire che Israele commette genocidio è una bestemmia».* Corriere Della Sera. https://www.corriere.it/esteri/24_maggio_22/liliana-segre-bestemmia-dire-che-israele-commette-genocidio-884e7dde-821b-4603-af69-8ba18a597xlk.shtml.

Fabian, E. (2025, January 2). *IDF suicide rate rises amid ongoing war and mass reservist call-ups.* https://www.timesofisrael.com/idf-suicide-rate-rises-amid-ongoing-war-and-mass-reservist-call-ups/.

Falla, D., Romera, E. M., & Ortega-Ruiz, R. (2021). Aggression, Moral Disengagement and Empathy. A Longitudinal Study Within the Interpersonal Dynamics of Bullying. *Frontiers in Psychology*, 12. https://doi.org/10.3389/fpsyg.2021.703468.

Farrar, F. W. (2018). *Families of Speech: Four Lectures Delivered Before the Royal Institution of Great Britain, in March 1869.* Forgotten Books.

Fatto Quotidiano. (2014, January 27). *Enzo Biagi intervista Primo Levi: 'Come nascono i lager? Facendo finta di nulla'.* http://www.ilfattoquotidiano.it/2014/01/27/enzo-biagi-intervista-primo-levi-come-nascono-i-lager-facendo-finta-di-nulla/858417/.

Fatto Quotidiano. (2025, January 15). *"A Gaza reazione giusta, non ci sono stati troppi morti": Le parole del ministro degli esteri israeliano da Bruno Vespa.* Ground News. https://www.ilfattoquotidiano.it/2025/01/15/ministro-esteri-israeliano-vespa-gaza-pochi-morti-reazione-giusta-news/7838653/.

Faulkner, N. (2018). 'Put Yourself in Their Shoes': Testing Empathy's Ability to Motivate Cosmopolitan Behavior. *Political Psychology, 39*(1), 217–228.

Fehr, E., & Gächter, S. (2002). Altruistic punishment in humans. *Nature, 415*(6868), 137–140. https://doi.org/10.1038/415137a.

Feminist Solidarity Network for Palestine. (2024, March 11). Here's what Pramila Patten's UN report on Oct 7 sexual violence actually said. *Mondoweiss.* https://mondoweiss.net/2024/03/heres-what-pramila-pattens-un-report-on-oct-7-sexual-violence-actually-said/.

Feng, Y., Warmenhoven, H., Wilson, A., Jin, Y., Chen, R., Wang, Y., & Hamer, K. (2022). The Identification With All Humanity (IWAH) scale: Its psychometric properties and associations with help-seeking during COVID-19. *Current Psychology (New Brunswick, N.j.)*, 1–13. https://doi.org/10.1007/s12144-022-03607-9.

Ferguson, N. (2011). *Civilization: The West and the Rest.* Allen Lane.

Ferrari, P. F., & Rizzolatti, G. (2014). Mirror neuron research: The past and the future. *Philosophical Transactions of the Royal Society B: Biological Sciences, 369*(1644), 20130169. https://doi.org/10.1098/rstb.2013.0169.

Finkelstein, N. G. (2003). *The Holocaust Industry: Reflections on the Exploitation of Jewish Suffering.* Verso.

Forensic Architecture. (2024). *A Cartography Of Genocide: Israel's Conduct In Gaza Since October 2023.* https://forensic-architecture.org//investigation/a-cartography-of-genocide.

Forensic Architecture. (2024a). *The Killing Of Hind Rajab.* https://forensic-architecture.org/investigation/the-killing-of-hind-rajab?fbclid=IwZXhobgNhZWoCMTEAARi g8Ag7wmgFOaGLLKpj_DZ19jwTqMENpqWgxvxe-jarf9A4Rih3ULkgPM8_aem_da4 wVMPb2KPwfuoF65i2PQ.

Forensic Architecture. (2024b, February 15). *Israeli Disinformation: Al-ahli Hospital.* https://forensic-architecture.org/investigation/israeli-disinformation-al-ahli-hospital.

Forensic Architecture. (2024c, October 17). *'when It Stopped Being A War': The Situated Testimony Of Dr Ghassan Abu-sittah ←.* https://forensic-architecture.org/investigation/when-it-stopped-being-a-war

Fowler, Z., Law, K. F., & Gaesser, B. (2021). Against Empathy Bias: The Moral Value of Equitable Empathy. *Psychological Science.* https://doi.org/10.1177/0956797620979965.

Frankel, J., & Bernstein, A. (2024, January 11). *Friendly fire may have killed their relatives on Oct. 7. These Israeli families want answers now.* AP News. https://apnews.com/article/israel-hamas-hostages-investigation-friendly-fire-3b6fdd4592957340b32a8ee71505b8e9.

Fraser, O. N., & Bugnyar, T. (2010). Do Ravens Show Consolation? Responses to Distressed Others. *PLOS ONE*, 5(5), e10605. https://doi.org/10.1371/journal.pone.0010605.

Freedom Channel. (2007, December 5). Milton Friedman on slavery and colonization Milton Friedman Speaks. *Freedom Channel, 2007.* https://freedomchannel.blogspot.com/2007/12/milton-friedman-on-slavery.html.

Freire, P. (1970). *Pedagogy of the Oppressed.* Seabury Press.

Frenkel, S. (2024, June 5). Israel Secretly Targets U.S. Lawmakers With Influence Campaign on Gaza War. *The New York Times.* https://www.nytimes.com/2024/06/05/technology/israel-campaign-gaza-social-media.html.

Friedman, A. (2024, October 15). *How a Rejected Course on the Israeli-Palestinian Conflict Rattled a Department.* The Chronicle of Higher Education. https://www.chronicle.com/article/a-curricular-clash-at-mit.

Fu, F., Tarnita, C. E., Christakis, N. A., Wang, L., Rand, D. G., & Nowak, M. A. (2012). Evolution of in-group favoritism. *Scientific Reports, 2,* 460. https://doi.org/10.1038/srep00460.

Galaria, I. (2024, February 16). *Opinion: I'm an American doctor who went to Gaza. What I saw wasn't war—it was annihilation.* Los Angeles Times. https://www.latimes.com/opinion/story/2024-02-16/rafah-gaza-hospitals-surgery-israel-bombing-ground-offensive-children.

Galeano E. (1973). *Open Veins of Latin America: Five Centuries of the Pillage of a Continent.* Monthly Review Press.

Gedeon, J. (2025, April 9). Pro-Israel group asks DoJ to investigate Ms Rachel over posts on Gaza children. *The Guardian.* https://www.theguardian.com/us-news/2025/apr/09/stop-antisemitism-ms-rachel-doj-investigation.

George, S. (1976). *How the Other Half Dies | Transnational Institute.* https://www.tni.org/en/publication/how-the-other-half-dies.

Gillman, A. S., Pérez-Stable, E. J., & Das, R. (2024). Advancing Health Disparities Science Through Social Epigenomics Research. *JAMA Network Open*, 7(7), e2428992. https://doi.org/10.1001/jamanetworkopen.2024.28992.

Goldberg, A. (2024, April 18). Yes, it is genocide. *The Palestine Project.* https://thepalestineproject.medium.com/yes-it-is-genocide-634a07ea27d4.

Goldenberg, T., & Frankel, J. (2024, May 22). *Israel-Hamas war: How 2 debunked accounts of sexual violence fueled global dispute | AP News.* https://apnews.com/article/israel-hamas-war-sexual-violence-zaka-ca7905bf9520b1e646f86d72cdf03244.

Goldstein, T. (2023, August 8). *2023 sets record for settlement construction and outpost legalization—watchdog.* https://www.timesofisrael.com/2023-sets-record-for-settlement-construction-and-outpost-legalization-watchdog/.

Gordon, A. (2023, November 10). *What Israelis Think of the War With Hamas.* TIME. https://time.com/6333781/israel-hamas-poll-palestine/.

Gove, M. (2025, January 8). *The IDF should be nominated for the Nobel Peace Prize.* The Jewish Chronicle. https://www.thejc.com/opinion/the-idf-should-be-nominated-for-the-nobel-peace-prize-xmppkld8.

Graham, J., Haidt, J., & Nosek, B. A. (2009). Liberals and conservatives rely on different sets of moral foundations. *Journal of Personality and Social Psychology, 96*(5), 1029–1046. https://doi.org/10.1037/a0015141.

Graham-Harrison, E. (2025, January 26). Trump's Gaza proposal rejected by allies and condemned as ethnic cleansing plan. *The Guardian.* https://www.theguardian.com/us-news/2025/jan/26/trump-resumes-sending-2000-pound-bombs-to-israel-undoing-biden-pause.

Graham-Harrison, E., & Abraham, Y. (2025, August 21). *Revealed: Israeli military's own data indicates civilian death rate of 83% in Gaza war. The Guardian.* https://www.theguardian.com/world/ng-interactive/2025/aug/21/revealed-israeli-militarys-own-data-indicates-civilian-death-rate-of-83-in-gaza-war.

Gray, T. S. (2016). Spencer, Herbert (1820–1903). In *Routledge Encyclopedia of Philosophy* (1st ed.). Routledge. https://doi.org/10.4324/9780415249126-DC076-1.

Griswold, E. (2024, March 21). The Children Who Lost Limbs in Gaza. *The New Yorker.* https://www.newyorker.com/news/dispatch/the-children-who-lost-limbs-in-gaza.

Gritten, D. (2024, April 5). *Gaza war: Where does Israel get its weapons?* https://www.bbc.com/news/world-middle-east-68737412.

Guarino, M. (2014, September 9). Professor fired for Israel criticism urges University of Illinois to reinstate him. *The Guardian.* https://www.theguardian.com/education/2014/sep/09/professor-israel-criticism-twitter-university-illinois.

Guillot, M., Draidi, M., Cetorelli, V., Silva, J. H. C. M. D., & Lubbad, I. (2025). Life expectancy losses in the Gaza Strip during the period October, 2023, to September, 2024. *The Lancet, 405*(10477), 478–485. https://doi.org/10.1016/S0140-6736(24)02810-1.

Guyénot, L. (2023, October 28). *Israel's Biblical Psychopathy—Réseau International.* https://en.reseauinternational.net/la-psychopathie-biblique-disrael/.

Haaretz. (2023, January 16). *Israel's Far-right Finance Minister Says He's 'A Fascist Homophobe' but 'Won't Stone Gays'—Israel News—Haaretz.com.* https://www.haaretz.com/israel-news/2023-01-16/ty-article/.premium/israels-far-right-finance-minister-im-a-fascist-homophobe-but-i-wont-stone-gays/00000185-b921-de59-a98f-ff7f47c70000.

Hajjaj, T. S. (2024, December 25). *'We will leave when the last Palestinian leaves': The defiant last stand of the doctors of Kamal Adwan Hospital.* Mondoweiss. https:// mondoweiss.net/2024/12/we-will-leave-when-the-last-palestinian-leaves-the-defi ant-last-stand-of-the-doctors-of-kamal-adwan-hospital/.

Hamas Elmasry, M. (2024). Images of the Israel-Gaza War on Instagram: A Content Analysis of Western Broadcast News Posts. *Journalism & Mass Communication Quarterly*, 10776990241287155. https://doi.org/10.1177/10776990241287155.

Hamer, K., Penczek, M., McFarland, S., Wlodarczyk, A., Łużniak-Piecha, M., Golińska, A., Cadena, L. M., Ibarra, M., Bertin, P., & Delouvée, S. (2021). Identification with all humanity—A test of the factorial structure and measurement invariance of the scale in five countries. *International Journal of Psychology*, *56*(1), 157–174. https://doi .org/10.1002/ijop.12678.

Hannah, D. (2023, October 30). *Why so many struggle to empathize with Israel— Washington Examiner.* https://www.washingtonexaminer.com/opinion/2576212 /why-so-many-struggle-to-empathize-with-israel/.

Harman, O. S. (Ed.). (2010). *The price of altruism: George Price and the search for the origins of kindness* (1. ed). W. W. Norton.

Harvey, F. (2022, February 28). *IPCC issues 'bleakest warning yet' on impacts of climate breakdown.* Climate Crisis | The Guardian. https://www.theguardian.com /environment/2022/feb/28/ipcc-issues-bleakest-warning-yet-impacts-climate -breakdown.

Hasson, N. (2024, December 5). A massive database of evidence, compiled by a historian, documents Israel's war crimes in Gaza. *Haaretz.* https://www.haaretz .com/israel-news/2024-12-05/ty-article-magazine/.highlight/massive-database-of -evidence-compiled-by-a-historian-details-israels-war-crimes-in-gaza/00000193 -979b-d408-a7d3-bfdbf1410000.

Hasson, N., & Rozovsky, L. (2023, December 4). *Hamas Committed Documented Atrocities. But a Few False Stories Feed the Deniers—Israel News.* Haaretz.Com. https://www.haaretz.com/israel-news/2023-12-04/ty-article-magazine/.premium /hamas-committed-documented-atrocities-but-a-few-false-stories-feed-the -deniers/0000018c-34f3-da74-afce-b5fbe24f0000.

Hedges, C. (2024a, October 25). Israel's War on Journalism [Substack newsletter]. *The Chris Hedges Report.* https://chrishedges.substack.com/p/israels-war-on-journal ism?utm_campaign=email-post-title&utm_medium=email.

Hedges, C. (2024b, December 10). *Letter to Refaat Alareer.* The Chris Hedges Report. https://chrishedges.substack.com/p/letter-to-refaat-alareer.

Hedges, C. (2024c, December 18). *Chris Hedges Report: Enduring the Trauma of Genocide.* Consortium News. https://consortiumnews.com/2024/12/18/chris-hedges -report-enduring-the-trauma-of-genocide/.

Hegel, G. W. F. (2001). *The philosophy of history* (*J. Sibree, Trans.*). Batoche Books.

Helmore, E. (2024, April 28). Echoes of Vietnam era as pro-Palestinian student protests roil US campuses. *The Guardian.* https://www.theguardian.com/world/2024/apr/28/us-student-protests-gaza-israel.

Helsinki Times. (2024, November 8). *Israeli hooligans' racist chants and provocation in Amsterdam spark international outrage.* https://www.helsinkitimes.fi/world-int/25763-israeli-hooligans-racist-chants-and-provocation-in-amsterdam-spark-international-outrage.html.

Hendrix, C. S., Koubi, V., Selby, J., Siddiqi, A., & von Uexkull, N. (2023). *Climate change and conflict.* https://doi.org/10.17863/CAM.101890.

Henrich, J. (2020a). *The WEIRDest People in the World: How the West Became Psychologically Peculiar and Particularly Prosperous.* Farrar, Straus and Giroux.

Henrich, J. (2020b). *The WEIRDest People in the World: How the West Became Psychologically Peculiar and Particularly Prosperous.* Farrar, Straus and Giroux.

Herman, E. S., & Chomsky, N. (2010). *Manufacturing Consent: The Political Economy of the Mass Media.* Random House.

Herrnstein, R. J., & Murray, C. A. (1994). *The Bell Curve: Intelligence and Class Structure in American Life.* Free Press.

Heyden, T. (2015, January 26). *The 10 greatest controversies of Winston Churchill's career—BBC News.* https://www.bbc.com/news/magazine-29701767.

Heym, N., Kibowski, F., Bloxsom, C. A. J., Blanchard, A., Harper, A., Wallace, L., Firth, J., & Sumich, A. (2021). The Dark Empath: Characterising dark traits in the presence of empathy. *Personality and Individual Differences, 169,* 110172. https://doi.org/10.1016/j.paid.2020.110172.

Horne, A. (2011). *A Savage War of Peace: Algeria 1954–1962.* New York Review of Books.

Huggel, C., Bouwer, L. M., Juhola, S., Mechler, R., Muccione, V., Orlove, B., & Wallimann-Helmer, I. (2022). The existential risk space of climate change. *Climatic Change, 174*(1), 8. https://doi.org/10.1007/s10584-022-03430-y.

Human Rights Watch. (2017, June 4). *Israel: 50 Years of Occupation Abuses.* https://www.hrw.org/news/2017/06/04/israel-50-years-occupation-abuses.

Human Rights Watch. (2022, June 14). *Gaza: Israel's 'Open-Air Prison' at 15.* https://www.hrw.org/news/2022/06/14/gaza-israels-open-air-prison-15.

Human Rights Watch. (2023, December 20). *Meta: Systemic Censorship of Palestine Content.* https://www.hrw.org/news/2023/12/20/meta-systemic-censorship-palestine-content.

Human Rights Watch. (2024a, July 17). *October 7 Crimes Against Humanity, War Crimes by Hamas-led Groups.* https://www.hrw.org/news/2024/07/17/october-7-crimes-against-humanity-war-crimes-hamas-led-groups.

Human Rights Watch. (2024c, December 19). *Israel's Crime of Extermination, Acts of Genocide in Gaza.* https://www.hrw.org/news/2024/12/19/israels-crime-extermination-acts-genocide-gaza.

Hume, D. (1987). Essays: Moral, Political, and Literary. Edited by Eugene F. Miller, revised edition. In *Essay XXI, 'Of National Characters'* (p. footnote 7, pp. 208–209). Indianapolis: Liberty Fund.

Huntington, S. P. (1996). *The Clash of Civilizations and the Remaking of World Order.* Simon & Schuster.

Huynh, B. Q., Chin, E. T., & Spiegel, P. B. (2024). No evidence of inflated mortality reporting from the Gaza Ministry of Health. *The Lancet, 403*(10421), 23–24. https://doi.org/10.1016/S0140-6736(23)02713-7.

ICJ. (2024). *Application of the Convention on the Prevention and Punishment of the Crime of Genocide in the Gaza Strip (South Africa v. Israel), Provisional Measures, Order of 26 January 2024.* International Court of Justice. https://www.icj-cij.org/sites/default/files/case-related/192/192-20240126-ord-01-00-en.pdf.

Idan, A. (2025, May 18). *Destroying Gaza 'With Love': Israel's New YogiNazis—Opinion—Haaretz.com.* https://www.haaretz.com/opinion/2025-05-18/ty-article-magazine/.premium/destroying-gaza-with-love-israels-new-yoginazis/00000196-d3c4-d048-a7d7-dbf6c43b0000.

IMEU Project. (2025, January). *IMEU Policy Project Post-Election Polling Shows Gaza Cost Harris Votes.* Institute for Middle East Understading Project. https://www.imeupolicyproject.org/postelection-polling.

Inbar, E., & Shamir, E. (2014, July 22). *Mowing the grass in Gaza—The Jerusalem Post.* https://www.jpost.com/Opinion/Columnists/Mowing-the-grass-in-Gaza-368516.

Inbari, M., & Bumin, K. M. (2024). Israeli Jewish Attitudes toward Core Religious Beliefs in God, the Election of Israel, Eschatology, and the Temple Mount—Statistical Analysis. *Religions, 15*(9), Article 9. https://doi.org/10.3390/rel15091076.

Innes, K. L. (2024, September 10). *Student camps for Gaza faced negative bias in The Globe and Mail, study shows ★ The Breach.* The Breach. https://breachmedia.ca/student-encampments-newspaper-coverage-biased-negative/.

International Association of Genocide Scholars. (2025, 31 August). Resolution on the situation in Gaza. https://genocidescholars.org/wp-content/uploads/2025/08/IAGS-Resolution-on-Gaza-FINAL.pdf.

IPCC. (2023). *IPCC, 2023: Climate Change 2023: Synthesis Report. Contribution of Working Groups I, II and III to the Sixth Assessment Report of the Intergovernmental Panel on Climate Change [Core Writing Team, H. Lee and J. Romero (eds.)]. IPCC, Geneva, Switzerland.* (First). Intergovernmental Panel on Climate Change (IPCC). https://doi.org/10.59327/IPCC/AR6-9789291691647.

Ismail, A. (2024, October 30). Yes, They're Really Voting for Jill Stein. *Slate.* https://slate.com/news-and-politics/2024/10/trump-harris-election-presidential-polls-israel.html.

Israel National News. (2023, April 26). *Ursula von der Leyen: 'Europe and Israel are bound to be friends and allies'.* Israel National News. https://www.israelnationalnews.com/news/370539.

Jackson, H. (2024). *The New York Times distorts the Palestinian struggle: A case study of anti-Palestinian bias in US news coverage of the First and Second Palestinian Intifadas.* 17(1), 116–135.

Jamaluddine, Z., Abukmail, H., Aly, S., Campbell, O. M. R., & Checchi, F. (2025). Traumatic injury mortality in the Gaza Strip from Oct 7, 2023, to June 30, 2024: A capture-recapture analysis. *Lancet (London, England)*, 405(10477), 469–477. https://doi.org/10.1016/S0140-6736(24)02678-3.

Jamaluddine, Z., Checchi, F., & Campbell, O. M. R. (2023). Excess mortality in Gaza: Oct 7–26, 2023. *The Lancet*, 402(10418), 2189–2190. https://doi.org/10.1016/S0140-6736(23)02640-5.

Jerusalem Post Staff. (2024, March 14). *Hamas's Gaza death toll is exaggerated or faked, statistics expert claims.* The Jerusalem Post | JPost.Com. https://www.jpost.com/israel-hamas-war/article-791838.

Johnson, A., & Ali, O. (2024, October 14). *A Study Reveals CNN and MSNBC's Glaring Gaza Double Standard.* https://www.thenation.com/article/society/cnn-msnbc-gaza-media-bias-study/.

Johnson, Adam, & Othman, A. (2024, January 9). *Coverage of Gaza War in the New York Times and Other Major Newspapers Heavily Favored Israel, Analysis Shows.* The Intercept. https://theintercept.com/2024/01/09/newspapers-israel-palestine-bias-new-york-times/.

Johnson, J. (2023, October 10). *Israeli Army Official Admits Gaza Bombing Campaign Is Focused on 'Damage and Not on Accuracy' | Common Dreams.* https://www.commondreams.org/news/israel-gaza-bombing.

JPPI. (2024, May 12). *Jewish People Policy Institute (JPPI) Israeli Society Index, May 2024.* https://jppi.org.il/en/16539-2/.

Kahane, G., Everett, J. A., Earp, B. D., Farias, M., & Savulescu, J. (2015). 'Utilitarian' judgments in sacrificial moral dilemmas do not reflect impartial concern for the greater good. *Cognition*, 134, 193. https://doi.org/10.1016/j.cognition.2014.10.005.

Kant, I. (2006). *Toward Perpetual Peace and Other Writings on Politics, Peace, and History.* Yale University Press.

Kant, I., & Gregor, M. J. (Eds.). (1996). The metaphysics of morals (1797). In *Practical Philosophy* (pp. 353–604). Cambridge University Press. https://doi.org/10.1017/CBO9780511813306.013.

Kapıkıran, N. A. (2023). Sources of Ethnocultural empathy: Personality, intergroup relations, affects. *Current Psychology*, 42(14), 11510–11528. https://doi.org/10.1007/s12144-021-02286-2.

Karni, D. (2024, August 6). *Israeli minister says it may be 'moral' to starve 2 million Gazans, but 'no one in the world would let us'.* CNN. https://www.cnn.com/2024/08/06/middleeast/israeli-minister-smotrich-starve-gazans-intl/index.html.

Karni, Y. (2015, August 15). *Bayit Yehudi MK: Gays control the media.* Ynetnews. https://www.ynetnews.com/articles/0,7340,L-4690957,00.html.

Kartal, A. G. (2024, April 26). *What we are seeing in Gaza is a 'repeat of Auschwitz,'* *says genocide expert.* https://www.aa.com.tr/en/middle-east/what-we-are-seeing -in-gaza-is-a-repeat-of-auschwitz-says-genocide-expert/3202869.

Kemp, L., Xu, C., Depledge, J., Ebi, K. L., Gibbins, G., Kohler, T. A., Rockström, J., Scheffer, M., Schellnhuber, H. J., Steffen, W., & Lenton, T. M. (2022). Climate Endgame: Exploring catastrophic climate change scenarios. *Proceedings of the National Academy of Sciences, 119*(34), e2108146119. https://doi.org/10.1073/pnas.2108146119.

Kennedy, K. (2024, September 9). *Florida Republican Randy Fine celebrates Israeli killing of American citizen in West Bank: 'Fire away!'* Tag 24. https://www.tag24.com /politics/politicians/florida-republican-randy-fine-celebrates-israeli-killing-of -american-citizen-in-west-bank-fire-away-3316388.

Khatib, R., McKee, M., & Yusuf, S. (2024). Counting the dead in Gaza: Difficult but essential. *The Lancet, 404*(10449), 237–238. https://doi.org/10.1016/S0140-6736(24)01169-3.

Khatsenkova, S. (2023, November 27). *Did Israeli children really sing about annihilating everyone in Gaza?* Euronews. https://www.euronews.com/my-europe/2023/11/27 /fact-check-did-israeli-children-really-sing-about-annihilating-everyone-in-gaza.

Koch, A., Brierley, C., Maslin, M. M., & Lewis, S. L. (2019). Earth system impacts of the European arrival and Great Dying in the Americas after 1492. *Quaternary Science Reviews, 207*, 13–36. https://doi.org/10.1016/j.quascirev.2018.12.004.

Kohn, M., & Reddy, K. (2024). Colonialism. In E. N. Zalta & U. Nodelman (Eds.), *The Stanford Encyclopedia of Philosophy* (Summer 2024). Metaphysics Research Lab, Stanford University. https://plato.stanford.edu/archives/sum2024/entries /colonialism/.

Krieger, N., Testa, C., Chen, J. T., Johnson, N., Watkins, S. H., Suderman, M., Simpkin, A. J., Tilling, K., Waterman, P. D., Coull, B. A., De Vivo, I., Smith, G. D., Roux, A. V. D., & Relton, C. (2023). Epigenetic aging & embodying injustice: US My Body My Story and Multi-Ethnic Atherosclerosis Study. *medRxiv*, 2023.12.13.23299930. https://doi .org/10.1101/2023.12.13.23299930.

Kristensen, & et al. (2025, March 26). Status of World Nuclear Forces. *Federation of American Scientists.* https://fas.org/initiative/status-world-nuclear-forces/.

Kropotkin, K. (2021). *Ethics: Origins and Development.* Black Rose Books Ltd.

Kropotkin, P. (2012). *Mutual Aid: A Factor of Evolution.* Courier Corporation.

Kull, S., Ramsay, C., & Lewis, E. (2003). Misperceptions, the Media, and the Iraq War. *Political Science Quarterly, 118*(4), 569–598.

La Repubblica. (2025, February 25). *Liliana Segre al presidio per i fratellini ebrei uccisi a Gaza: "Essere qui è la risposta a tutto".* la Repubblica. https://milano.repubblica .it/cronaca/2025/02/25/news/liliana_segre_presidio_fratellini_ebrei_uccisi_a _gaza_bibas_essere_qui_e_risposta_tutto-424027696/.

Laband, J. (2015). Elizabethvan Heyningen. The Concentration Camps of the Anglo-Boer War: A Social History. *The American Historical Review, 120*(2), 760. https://doi. org/10.1093/ahr/120.2.760.

Lacy, A. (2024, November 6). *Trump Didn't Win Pennsylvania. Kamala Harris Lost It.* The Intercept. https://theintercept.com/2024/11/06/trump-harris-pennsylvania-swing-state-election-results/.

Lanard, N. (2023, November 3). The dangerous history behind Netanyahu's Amalek rhetoric. *Mother Jones.* https://www.motherjones.com/politics/2023/11/benjamin-netanyahu-amalek-israel-palestine-gaza-saul-samuel-old-testament/.

Landy, D., Lentin, R., & McCarthy, C. (2020). *Enforcing Silence: Academic Freedom, Palestine and the Criticism of Israel.* Zed Books Ltd.

Laville, S. (2025, January 16). *Global economy could face 50% loss in GDP between 2070 and 2090 from climate shocks, say actuaries | Climate crisis | The Guardian.* https://www.theguardian.com/environment/2025/jan/16/economic-growth-could-fall-50-over-20-years-from-climate-shocks-say-actuaries.

Law for Palestine. (2024, January 4). Law for Palestine Releases Database with 500+ Instances of Israeli Incitement to Genocide—Continuously Updated. *Law for Palestine.* https://law4palestine.org/law-for-palestine-releases-database-with-500-instances-of-israeli-incitement-to-genocide-continuously-updated/.

Law, K., Campbell, D., & Gaesser, B. (2022). Biased Benevolence: The Perceived Morality of Effective Altruism Across Social Distance. *Personality & Social Psychology Bulletin, 48*(3). https://doi.org/10.1177/01461672211002773.

Le Nevé, S. (2024, September 4). *Un enseignant de la Toulouse School of Economics suspendu après avoir appelé au boycott d'Israël.* https://www.lemonde.fr/societe/article/2024/09/04/un-enseignant-de-la-toulouse-school-of-economics-suspendu-apres-avoir-appele-au-boycott-d-israel_6304295_3224.html.

Lemkin Institute. (2025, May 28). *Statement Calling Out US & Israeli Propaganda.* Lemkin Institute for Genocide Prevention. https://www.lemkininstitute.com/statements-new-page/statement-calling-out-us-%26-israeli-propaganda-.

Lenharo, M. (2023). Morality is declining, right? Scientists say that idea is an illusion. *Nature, 618*(7965), 441–442. https://doi.org/10.1038/d41586-023-01848-7.

Li, X.-C., Zhang, Y.-Y., Zhang, Q.-Y., Liu, J.-S., Ran, J.-J., Han, L.-F., & Zhang, X.-X. (2024). Global burden of viral infectious diseases of poverty based on Global Burden of Diseases Study 2021. *Infectious Diseases of Poverty, 13,* 71. https://doi.org/10.1186/s40249-024-01234-z.

Liddell, J. (2024, May 13). *Lindsey Graham suggests Israel should nuke Gaza.* The Independent. https://www.independent.co.uk/news/world/americas/us-politics/lindsey-graham-gaza-israel-hiroshima-b2544204.html.

Liu, X., Yang, X., & Wu, Z. (2021). To Punish or to Restore: How Children Evaluate Victims' Responses to Immorality. *Frontiers in Psychology, 12,* 696160. https://doi.org/10.3389/fpsyg.2021.696160.

Locke, J. (2016). *Two Treatises of Government.* CreateSpace Independent Publishing Platform.

Luberto, C. M., Shinday, N., Song, R., Philpotts, L. L., Park, E. R., Fricchione, G. L., & Yeh, G. Y. (2018). A Systematic Review and Meta-analysis of the Effects of Meditation on Empathy, Compassion, and Prosocial Behaviors. *Mindfulness*, *9*(3), 708–724. https://doi.org/10.1007/s12671-017-0841-8.

Luhnow, C. K.-L. and D. (2024, January 29). *Intelligence Reveals Details of U.N. Agency Staff's Links to Oct. 7 Attack*. wsj. https://www.wsj.com/world/middle-east/at-least-12-u-n-agency-employees-involved-in-oct-7-attacks-intelligence-reports-say-a7de8f36.

Lukiv, & Gritten, D. (2025, March 31). *Gaza: Red Cross outraged over killing of medics by Israeli forces*. https://www.bbc.com/news/articles/crkxm1rg6k1o.

Maad, A., Audureau, W., & Forey, S. (2024, April 3). *'40 beheaded babies': Deconstructing the rumor at the heart of the information battle between Israel and Hamas*. https://www.lemonde.fr/en/les-decodeurs/article/2024/04/03/40-beheaded-babies-the-itinerary-of-a-rumor-at-the-heart-of-the-information-battle-between-israel-and-hamas_6667274_8.html.

MacEoin, D. (2024, June 3). *Denis MacEoin: "Cari studenti, Israele non è un regime"*. la Repubblica. https://www.repubblica.it/cultura/2024/06/03/news/denis_maceoin_israele_lettera_studenti_edimburgo-423157522/.

Mahase, E. (2023). *Gaza: "Unprecedented" bombing of hospital leaves hundreds dead and injured*. https://doi.org/10.1136/bmj.p2423.

Mahomed, S. (2023). When sanctuaries of humanity turn into corridors of horror: The destruction of healthcare in Gaza. *South African Journal of Bioethics and Law*, *16*(3), 77–79. https://doi.org/10.7196/SAJBL.2023.v16i3.1732.

Malik, N. (2025, January 27). Goodbye to the lost children of Gaza. You were loved, you are remembered, you did not deserve it. *The Guardian*. https://www.theguardian.com/commentisfree/2025/jan/27/children-gaza-loved-remembered-innocent-mourn.

Mallik, S. (2023). Colonial Biopolitics and the Great Bengal Famine of 1943. *Geojournal*, *88*(3), 3205–3221. https://doi.org/10.1007/s10708-022-10803-4.

Mancke E, Shammas C. (2005). *The Creation of the British Atlantic world*. JHU Press.

Margalit, R. (2023, February 20). Itamar Ben-Gvir, Israel's Minister of Chaos. *The New Yorker*. https://www.newyorker.com/magazine/2023/02/27/itamar-ben-gvir-israels-minister-of-chaos.

Marques, N. (2024, June 5). Healthcare workers will protest the American Medical Association's silence in the face of genocide. *Peoples Dispatch*. https://peoplesdispatch.org/2024/06/05/healthcare-workers-will-protest-the-american-medical-associations-silence-in-the-face-of-genocide/.

Masih, N. (2023, December 13). *Here are countries that voted for and against the U.N. Gaza cease-fire resolution*. The Washington Post. https://www.washingtonpost.com/world/2023/12/13/un-vote-gaza-ceasefire-countries-against/?utm_source=chatgpt.com.

Massad, J. (2024, August 30). *Why Israel's genocide in Gaza is a western war on the Palestinian people | Middle East Eye.* https://www.middleeasteye.net/opinion /israel-genocide-gaza-western-war-palestinian-people.

Maté, G. (2022). *The Myth of Normal: Trauma, Illness, and Healing in a Toxic Culture.* Penguin.

Mazza, P. (2024, May 17). On the Campuses and Around the World, a Revolution of Empathy. *CounterPunch.Org.* https://www.counterpunch.org/2024/05/17/on-the -campuses-and-around-the-world-a-revolution-of-empathy/.

McAuliffe, K., Blake, P. R., Steinbeis, N., & Warneken, F. (2017). The developmental foundations of human fairness. *Nature Human Behaviour, 1*(2), 1–9. https://doi .org/10.1038/s41562-016-0042.

McDoom, O. S. (2024). Expert Commentary, the Israeli-Palestinian Conflict, and the Question of Genocide: Prosemitic Bias within a Scholarly Community? *Journal of Genocide Research, 0*(0), 1–9. https://doi.org/10.1080/14623528.2024.2346403.

McEvoy, J. (2024, June 5). *Israel lobby funded a quarter of British MPs.* Declassified UK. https://www.declassifieduk.org/israel-lobby-funded-a-quarter-of-british-mps/ https://www.youtube.com/watch?v=IETxE7VErPA.

McFarland, S. G. (2021). *Heroes of Human Rights: Stories of Women and Men Who Created Human Rights.* Cognella, Incorporated.

McFarland, S., Webb, M., & Brown, D. (2012). All humanity is my ingroup: A measure and studies of identification with all humanity. *Journal of Personality and Social Psychology, 103*(5), 830–853. https://doi.org/10.1037/a0028724.

McGreal, C. (2022, December 15). 'I became more and more violent': Shocking testimonies of abuse from IDF veterans. *The Guardian.* https://www.theguardian.com /artanddesign/2022/dec/15/idf-exhibition-breaking-the-silence.

McGreal, C. (2023, October 16). The language being used to describe Palestinians is genocidal. *The Guardian.* https://www.theguardian.com/commentisfree/2023/oct /16/the-language-being-used-to-describe-palestinians-is-genocidal.

McGreal, C. (2024a, February 4). CNN staff say network's pro-Israel slant amounts to 'journalistic malpractice'. *The Guardian.* https://www.theguardian.com/media /2024/feb/04/cnn-staff-pro-israel-bias.

McGreal, C. (2024b, April 2). 'Not a normal war': Doctors say children have been targeted by Israeli snipers in Gaza. *The Guardian.* https://www.theguardian.com /world/2024/apr/02/gaza-palestinian-children-killed-idf-israel-war.

McKernan, B. (2023, October 18). *'They believed it was safe': Death toll rising after blast at Gaza hospital | Israel-Gaza war | The Guardian.* https://www.theguardian.com /world/2023/oct/18/they-believed-it-was-safe-death-toll-rising-blast-gaza-hospital.

McKie, R. (2022, July 30). 'Soon the world will be unrecognisable': Is it still possible to prevent total climate meltdown? *The Observer.* https://www.theguardian .com/environment/2022/jul/30/total-climate-meltdown-inevitable-heatwaves -global-catastrophe.

McLaughlin, K. A., Sheridan, M. A., & Lambert, H. K. (2014). Childhood adversity and neural development: Deprivation and threat as distinct dimensions of early experience. *Neuroscience and Biobehavioral Reviews, 47,* 578–591. https://doi.org/10.1016/j.neubiorev.2014.10.012.

McLaughlin, R. (2024, December 15). *'Do you value life, SJSU?': Longtime professor resigns over San Jose State University support for Gaza genocide.* Mondoweiss. https://mondoweiss.net/2024/12/do-you-value-life-sjsu-longtime-professor-resigns-over-san-jose-state-university-support-for-gaza-genocide/.

McLoughlin, P. (2025, February 20). *Lausanne ends Joseph Daher contract 'over pro-Palestine stance'.* https://www.newarab.com/; The New Arab. https://www.newarab.com/news/lausanne-ends-joseph-daher-contract-over-pro-palestine-stance.

Mechanic, M. (2021, April 4). Research Proves It: There's No Such Thing as Noblesse Oblige. *The Atlantic.* https://www.theatlantic.com/ideas/archive/2021/04/does-wealth-rob-brain-compassion/618496/.

Media Lens. (2021, February 23). 20 Years Of Media Lens: A Selection Of Remarkable Replies From Journalists. *Media Lens.* https://www.medialens.org/2021/20-years-of-media-lens-a-selection-of-remarkable-replies-from-journalists/.

Media Matters. (2023, December 18). *Michelle Bachmann to Charlie Kirk: 'It's time that Gaza ends. The 2 million people who live there, they are clever assassins. They need to be removed from that land.'* Media Matters. https://www.mediamatters.org/charlie-kirk/michelle-bachmann-charlie-kirk-its-time-gaza-ends-2-million-people-who-live-there-they.

Media Mondo. (2024, August 13). *Interview Rabbi Yaakov Shapiro—The Palestine Pod.* https://media-mondo.com/en/videos-en/interview-rabbi-yaakov-shapiro-the-palestine-pod/.

Middle East Eye. (2023a, July 27). *Ex-Mossad chief says Netanyahu government worse than Ku Klux Klan.* Middle East Eye. https://www.middleeasteye.net/news/israel-ex-mossad-chief-says-netanyahu-government-worse-kkk.

Middle East Eye. (2023b, November 5). *Israel-Palestine war: 100 Israeli doctors call for Gaza hospitals to be bombed.* Middle East Eye. https://www.middleeasteye.net/news/israel-palestine-war-doctors-call-gaza-hospitals-bombed.

Middle East Eye. (2025, May 14). *On Tuesday, Knesset member Michal Waldiger of the Religious Zionist Party responded to a speech by Knesset member Ahmad Tibi of the Hadash-Ta'al party, in which he condemned the killing of children in Gaza. During her remarks, a heated exchange broke out between Waldiger and* https://t.co/htMEPrjyYn [Tweet]. Twitter. https://x.com/MiddleEastEye/status/1922738119109234968.

Middle East Monitor. (2023a, October 9). *Israel MK calls for a second Nakba in Gaza.* https://www.middleeastmonitor.com/20231009-israel-mk-calls-for-a-second-nakba-in-gaza/.

Middle East Monitor. (2023b, October 10). *Israel MK says 'use Doomsday weapons' against Gaza*. Middle East Monitor. https://www.middleeastmonitor.com/20231010 -israel-mk-says-use-doomsday-weapons-against-gaza/.

Middle East Monitor. (2024a, March 8). *UN rights expert: 'Israel's starvation of Gaza is genocide'*. https://www.middleeastmonitor.com/20240308-un-rights-expert-israels -starvation-of-gaza-is-genocide/?fbclid=IwZXh0bgNhZWoCMTAAAR4rbhP _ObmUzDw5KmlyICOfyHDWDFkoTmAKmwQo7IsbgacgBdE2ZkXfrGBGxQ_aem _wVQsgq-lv2TP6BrcvJXLPA.

Middle East Monitor. (2024b, April 29). *Ben-Gvir calls on soldiers to 'kill' not arrest Palestinians who surrender in Gaza*. Middle East Monitor. https://www.middleeast monitor.com/20240429-ben-gvir-calls-on-soldiers-to-kill-not-arrest-palestinians -who-surrender-in-gaza/.

Middle East Monitor. (2024c, May 28). *European Commission president accused of complicity in Israel's war crimes at ICC*. Middle East Monitor. https://www.mid dleeastmonitor.com/20240528-european-commission-president-accused-of-com plicity-in-israels-war-crimes-at-icc/.

Middle East Monitor. (2024d, October 17). *Gabor Mate: 'It's like watching Auschwitz on TikTok'*. https://www.middleeastmonitor.com/20241017-gabor-mate-its-like-watching -auschwitz-on-tiktok/.

Middle East Monitor. (2025a). *A civilisation has been wiped out in Gaza, says Trump, admitting Israeli genocide—Middle East Monitor*. https://www.middleeastmonitor .com/20250212-a-civilisation-has-been-wiped-out-in-gaza-says-trump-admitting -israeli-genocide/.

Middle East Monitor. (2025b, January 6). *No rape allegations filed from 7 October, reveals Israeli prosecutor*. https://www.middleeastmonitor.com/20250106-no-rape -allegations-filed-from-7-october-reaveals-israeli-prosecutor/https://www.ynet .co.il/news/article/yokra14200599.

Middle East Monitor. (2025c, February 26). *'The silence for Gazan kids killed would last 300 hours'*. https://www.middleeastmonitor.com/20250226-the-silence-for-gazan -kids-killed-would-last-300-hours/.

Milgram, S. (1963). Behavioral Study of obedience. *The Journal of Abnormal and Social Psychology*, *67*(4), 371–378. https://doi.org/10.1037/h0040525.

Miller, A. (1990). *For Your Own Good: Hidden Cruelty in Child-Rearing and the Roots of Violence*. Farrar, Straus and Giroux.

Misgav, U. (2024, May 23). *Israel Should Have Arrested Arch-criminal Netanyahu— Opinion—Haaretz.com*. https://www.haaretz.com/opinion/2024-05-23/ty-article -opinion/.premium/israel-should-have-arrested-arch-criminal-netanyahu/00000 18f-a1a3-d639-a5bf-f3fbbf660000.

Mitrovica. (2023). *Sociopaths in suits | Israel-Palestine conflict | Al Jazeera*. https://www .aljazeera.com/opinions/2023/11/2/sociopaths-in-suits.

Mogahed, D. (2024, September 17). Winning Muslim Votes: A Policy Priority Analysis in Swing States. *Institute for Social Policy and Understanding (ISPU)*. https://ispu.org/winning-muslim-votes-1/.

Moll, J., Krueger, F., Zahn, R., Pardini, M., de Oliveira-Souza, R., & Grafman, J. (2006). Human fronto—mesolimbic networks guide decisions about charitable donation. *Proceedings of the National Academy of Sciences of the United States of America, 103*(42), 15623–15628. https://doi.org/10.1073/pnas.0604475103.

Monroe, K. (2006, July 23). *The Hand of Compassion | Princeton University Press*. https://press.princeton.edu/books/paperback/9780691127736/the-hand-of-compassion.

Monroe, R. (2017, September 14). *New Climate Risk Classification Created to Account for Potential 'Existential' Threats | Scripps Institution of Oceanography*. https://scripps.ucsd.edu/news/new-climate-risk-classification-created-account-potential-existential-threats.

Montero, E., & Lowes, S. (2019, August 14). *King Leopold's ghost: The legacy of labour coercion in the DRC*. CEPR. https://cepr.org/voxeu/columns/king-leopolds-ghost-legacy-labour-coercion-drc.

Mordechai, L. (2024). *Bearing Witness to the Israel-Gaza War*. https://www.academia.edu/112967602/Bearing_Witness_to_the_Israel_Gaza_War_updated_to_15_April_2024_.

Moussa, E. (2018). Victimhood, Selective Empathy, And Unique Historical Causality in Zionist Thinking. *Unpublished Manuscript*. https://www.researchgate.net/publication/328783494_Victimhood_Selective_Empathy_and_Unique_Historical__Causality_in_Zionist_Thinking.

Mulla, I. (2024, December 2). *UK could face legal action for 'genocide complicity', SNP politician Chris Law says*. Middle East Eye. https://www.middleeasteye.net/news/mp-warns-uk-ministers-significant-risk-legal-action-genocide-complicity.

Myers, S. L., & Frenkel, S. (2023, March 3). *In a Worldwide War of Words, Russia, China and Iran Back Hamas*. The New York Times. https://www.nytimes.com/2023/11/03/technology/israel-hamas-information-war.html.

NDTV. (2025, February 9). *Ex Israel Army Chief Admits Using Hannibal Directive Against Own Soldiers*. Www.Ndtv.Com. https://www.ndtv.com/world-news/yoav-gallant-admits-to-authorising-hannibal-directive-during-october-7-attack-7663931.

Nugent, A. (2022, February 27). *CBS News foreign correspondent apologises for saying Ukraine is more 'civilised' than Iraq and Afghanistan | The Independent*. https://www.independent.co.uk/arts-entertainment/tv/news/charlie-dagata-cbs-apology-ukraine-iraq-b2024265.html.

Nur, I. (2023, November 21). *Video Purportedly Shows Israeli Children Singing 'We Will Annihilate Everyone' in Gaza*. Yahoo News. https://www.yahoo.com/news/actual-video-shows-israeli-children-002000401.html.

O'Brien & Gould J. (2024, September 9). *US has seen no evidence that Israel has committed genocide, Austin says.* POLITICO. https://www.politico.com/news/2024/04/09/us-has-seen-no-evidence-that-israel-has-committed-genocide-austin-says-00151241.

Office of the High Commissioner for Human Rights. (2025, March 21). *Ukraine: Devastating impact of hostilities on children's rights.* https://www.ohchr.org/en/press-releases/2025/03/ukraine-devastating-impact-hostilities-childrens-rights.

Office of the United Nations High Commissioner for Human Rights. (2025, July 15). *Israel must stop killings and home demolitions in occupied West Bank.* https://www.ohchr.org/en/press-briefing-notes/2025/07/israel-must-stop-killings-and-home-demolitions-occupied-west-bank.

Ofir, J. (2019, January 18). Israeli historian Benny Morris doubles down on his advocacy for ethnic cleansing. *Mondoweiss.* https://mondoweiss.net/2019/01/historian-advocacy-cleasning/.

Ofir, J. (2023, May 13). *Liberal Israeli icon calls to raze Gaza neighborhoods.* Mondoweiss. https://mondoweiss.net/2023/05/liberal-israeli-icon-calls-to-raze-gaza-neighborhoods/.

Ofir, J. (2024, August 19). *65% of Israeli Jews oppose criminal prosecution for soldiers suspected of raping Palestinian detainees.* Mondoweiss. https://mondoweiss.net/2024/08/65-of-israeli-jews-oppose-criminal-prosecution-for-soldiers-suspected-of-raping-palestinian-detainees/.

OHCHR. (2023a, July 10). *Special Rapporteur Says Israel's Unlawful Carceral Practices in the Occupied Palestinian Territory Are Tantamount to International Crimes and Have Turned it into an Open-Air Prison.* OHCHR. https://www.ohchr.org/en/news/2023/07/special-rapporteur-says-israels-unlawful-carceral-practices-occupied-palestinian.

OHCHR. (2023b, October 14). *UN expert warns of new instance of mass ethnic cleansing of Palestinians, calls for immediate ceasefire.* OHCHR. https://www.ohchr.org/en/press-releases/2023/10/un-expert-warns-new-instance-mass-ethnic-cleansing-palestinians-calls.

OHCHR. (2024a, February 1). *Gaza: UN experts condemn killing and silencing of journalists.* OHCHR. https://www.ohchr.org/en/press-releases/2024/02/gaza-un-experts-condemn-killing-and-silencing-journalists.

OHCHR. (2024b, July 31). *Thematic Report—Detention in the context of the escalation of hostilities in Gaza (October 2023–June 2024).* OHCHR. https://www.ohchr.org/en/documents/reports/thematic-report-detention-context-escalation-hostilities-gaza-october-2023-june.

OHCHR. (2024c, August 5). *Israel's escalating use of torture against Palestinians in custody a preventable crime against humanity: UN experts.* OHCHR. https://www.ohchr.org/en/press-releases/2024/08/israels-escalating-use-torture-against-palestinians-custody-preventable.

OHCHR. (2025a, January 2). *UN experts horrified at blatant disregard for health rights in Gaza following deadly raid on Kamal Adwan hospital.* OHCHR. https://www

.ohchr.org/en/press-releases/2025/01/un-experts-horrified-blatant-disregard-health-rights-gaza-following-deadly.

OHCHR. (2025b, March 13). *"More than a human can bear": Israel's systematic use of sexual, reproductive and other forms of gender-based violence since October 2023.* OHCHR. https://www.ohchr.org/en/press-releases/2025/03/more-human-can-bear-israels-systematic-use-sexual-reproductive-and-other.

OHCHR. (2025c, May 7). *End unfolding genocide or watch it end life in Gaza: UN experts say States face defining choice.* https://www.ohchr.org/en/press-releases/2025/05/end-unfolding-genocide-or-watch-it-end-life-gaza-un-experts-say-states-face.

OHCHR. (2025d, June 10). *Israeli attacks on educational, religious and cultural sites in the Occupied Palestinian Territory amount to war crimes and the crime against humanity of extermination, UN Commission says.* OHCHR. https://www.ohchr.org/en/press-releases/2025/06/israeli-attacks-educational-religious-and-cultural-sites-occupied.

Olson, K. R., & Spelke, E. S. (2008). Foundations of cooperation in young children. *Cognition, 108*(1), 222–231. https://doi.org/10.1016/j.cognition.2007.12.003.

Oren, Z. (2024, August 7). *Three Israeli army refusers: 'We will not participate in genocide'.* +972 Magazine. https://www.972mag.com/israeli-army-refuseniks-moav-mueller-greenberg/.

Osman, N. (2024, March 14). *Israeli academic resigns from Hebrew University after Palestinian colleague suspended.* Middle East Eye. https://www.middleeasteye.net/news/war-gaza-professor-yuri-pines-resigns-hebrew-university-following-suspension-palestinian.

Otu-Larbi, F., Neimark, B., Bigger, P., & Cottrell, L. (2024). *A Multitemporal Snapshot of Greenhouse Gas Emissions from the Israel-Gaza Conflicta* (Queen Mary University of London). https://www.qmul.ac.uk/sbm/media/sbm/documents/Gaza_Carbon_Emissions.pdf.

Pagadala, M. S., & Nichols, N. (2025). Responses to the Gaza-Israel Conflict by Specialty Medical Societies. *JAMA Network Open, 8*(4), e254662. https://doi.org/10.1001/jamanetworkopen.2025.4662.

Pappe', I. (2007). *The Ethnic Cleansing of Palestine.* Oneworld Publications.

Pappe', I. (2017). *Ten Myths About Israel.* Verso Books.

Patel, K. (2025, March 5). *I'm a former BBC newsreader—Gaza is the reason I resigned.* The Independent. https://www.independent.co.uk/voices/bbc-gaza-documentary-tim-davie-israel-b2708547.html.

Patnaik. (2009). *Wired to Care: How Companies Prosper When They Create Widespread Empathy.* Pearson Education India.

PAX. (2024, June). The companies arming Israel and their financiers. *PAX.* https://paxforpeace.nl/publications/the-companies-arming-israel-and-their-financiers/.

PCPSR. (2024, July). *The Palestine/Israel Pulse, a Joint Poll: Press Release* [Palestinian Center for Policy and Survey Research]. https://www.pcpsr.org/en/node/989.

Pedwell C. (2016). De-Colonizing Empathy: Thinking Affect Transnationally. *Samyukta: A Journal of Gender and Culture*, 1(1). https://doi.org/10.53007/SJGC.2016.V1.I1.51

Peterson, J. (2023, December 14). Pro-Hamas protesters are sanctimonious psychopaths. *The Telegraph*. https://www.telegraph.co.uk/news/2023/12/14/jordan-peterson-pro-hamas-protesters-sanctimonious-psychopa/.

Pfaff, D. W. (2007). *The Neuroscience of Fair Play: Why We (usually) Follow the Golden Rule*. Dana Press.

Pfaff, D. W. (2014). *The Altruistic Brain: How We Are Naturally Good*. Oxford University Press.

Phillips, S. T., & Ziller, R. C. (1997). Toward a theory and measure of the nature of nonprejudice. *Journal of Personality and Social Psychology*, 72(2), 420–434. https://doi.org/10.1037/0022-3514.72.2.420.

Philo, G. (2002, April 16). Missing in action. *The Guardian*. https://www.theguardian.com/world/2002/apr/16/israelandthepalestinians.media.

Philp, C., & Weiniger, G. (2024, June 7). *Israel says Hamas weaponised rape. Does the evidence add up?* The Times. https://www.thetimes.com/magazines/the-times-magazine/article/israel-hamas-rape-investigation-evidence-october-7-6kzphszsj.

Plagakis, S., & Torreon, B. S. (2023, June 7). *Instances of Use of United States Armed Forces Abroad, 1798–2023*. https://www.congress.gov/crs-product/R42738.

Pletka & Soleimany. (2024). *Israel Is Not Committing 'Genocide' in Gaza*. https://www.aei.org/op-eds/israel-is-not-committing-genocide-in-gaza/.

Pörtner, H.-O., Roberts, D. C., Tignor, M. M. B., Poloczanska, E. S., Mintenbeck, K., Alegría, A., Craig, M., Langsdorf, S., Löschke, S., Möller, V., Okem, A., & Rama, B. (Eds.). (2022). *Climate Change 2022: Impacts, Adaptation and Vulnerability. Contribution of Working Group II to the Sixth Assessment Report of the Intergovernmental Panel on Climate Change*.

Prime Minister's Office. (2023, November 23). *PM Netanyahu Visits ZAKA*. Www.Gov.Il. https://www.gov.il/en/pages/event-zaka231123.

Prosinger, J. (2024, November 29). Human Rights Expert: 'A Strong Case that Israel's Response Constitutes the Crime of Genocide'. *Der Spiegel*. https://www.spiegel.de/international/world/interview-with-human-rights-expert-william-schabas-a-strong-case-that-israels-response-constitutes-the-crime-of-genocide-a-da7e4524-ab3b-40e4-b409-f8fca9c081b8.

Quigley, J. B. (2024, March 14). Legal Standard for Genocide Intent: An Uphill Climb for Israel in Gaza Suit. *EJIL: Talk!* https://www.ejiltalk.org/legal-standard-for-genocide-intent-an-uphill-climb-for-israel-in-gaza-suit/.

Quinn, R. (2024, June 3). *DePaul Adjunct Ousted for Optional Gaza Assignment*. Inside Higher Ed. https://www.insidehighered.com/news/faculty-issues/academic-freedom/2024/06/03/depaul-adjunct-ousted-optional-gaza-assignment.

RAI (Director). (2024, November 5). *Puntata del 05/11/2024 ore 14:50 | TGR Leonardo, Rubrica del TGR* [Video recording]. Radio Televisione Italiana 3. https://www

.rainews.it/tgr/rubriche/leonardo/video/2024/11/TGR-Leonardo-del-05112024 -6e32ea9c-a797-46cb-9ab7-30df8916e7a9.html.

RAI 3 (Director). (2023, October 10). *TV program Restart* [Video recording]. https:// www.raiplay.it/video/2023/10/Restart---Puntata-del-10102023-dbde6e46-c7a8 -4f8b-abf7-dedd3ee46f34.html.

Rama, P. (2024, April 3). Wisconsin's 'uninstructed' voters send Biden a strong message on the war in Gaza. *NPR*. https://www.npr.org/2024/04/03/1242424035 /wisconsin-uninstructed-uncommitted-biden-gaza.

Range, F., Horn, L., Viranyi, Z., & Huber, L. (2009). The absence of reward induces inequity aversion in dogs. *Proceedings of the National Academy of Sciences, 106*(1), 340–345. https://doi.org/10.1073/pnas.0810957105.

Rapaport, N. (2025, May 24). *Nearly half of Israelis support army killing all Palestinians in Gaza, poll finds*. Middle East Eye. https://www.middleeasteye.net/news /majority-israelis-support-expulsion-palestinians-gaza-poll.

Rascius, B. (2024, March 22). *Have more Palestinians or Israelis died in war? Half of Americans don't know, poll says*. Miami Herald. https://www.miamiherald.com /news/nation-world/national/article286996080.html.

Rascius, B. (2025, January 15). *Why did Harris lose some Biden 2020 voters? Poll finds Gaza war was the top issue*. Miami Herald. https://www.miamiherald.com/news /nation-world/national/article298600563.html.

Rashid, H. (2024, May 28). *Nikki Haley Signs Her Name on Israeli Bombs—Alongside Sick Message*. Yahoo News. https://www.yahoo.com/news/nikki-haley-signs -her-name-181359506.html.

Reid, S. (2024, February 21). *US Congressman Andy Ogles stirs outrage with Gaza comment—'Kill them all'*. Al Jazeera. https://www.aljazeera.com/news/2024/2/21 /us-congressman-andy-ogles-stirs-outrage-with-gaza-comment-kill-them-all.

Reiff, B. (2024a, April 5). *Israeli teen jailed for refusing draft: 'I'm willing to pay a price for my principles'*. +972 Magazine. https://www.972mag.com/ben-arad-conscientious -objector-israeli-army/.

Reiff, B. (2024b, November 11). *The not-so-secret history of Netanyahu's support for Hamas*. +972 Magazine. https://www.972mag.com/netanyahu-hamas-october-7 -adam-raz/.

Reporters Without Borders. (2024). *2024 World Press Freedom Index—journalism under political pressure*. Reporters Without Borders. https://rsf.org/en/2024-world -press-freedom-index-journalism-under-political-pressure.

Reuters. (2023, November 17). *Canada, Britain and main EU countries join Myanmar genocide case | Reuters*. https://www.reuters.com/world/canada-britain-main-eu -countries-join-myanmar-genocide-case-2023-11-17/.

Reuters. (2024a, March 8). *UNRWA report says Israel coerced some agency employees to falsely admit Hamas links*. https://www.reuters.com/world/middle-east

/unrwa-report-says-israel-coerced-some-agency-employees-falsely-admit-hamas
-links-2024-03-08/.

Reuters. (2024b, March 14). *Italy arms exports to Israel continued despite block, minister says.* https://www.reuters.com/world/europe/italy-arms-exports-israel-continued
-despite-block-minister-says-2024-03-14/.

Reysen, S., & Hackett, J. (2016). Further Examination of the Factor Structure and Validity of the Identification with All Humanity Scale. *Current Psychology, 35*(4), 711–719. https://doi.org/10.1007/s12144-015-9341-y.

Richardson, K., Steffen, W., Lucht, W., Bendtsen, J., Cornell, S. E., Donges, J. F., Drüke, M., Fetzer, I., Bala, G., von Bloh, W., Feulner, G., Fiedler, S., Gerten, D., Gleeson, T., Hofmann, M., Huiskamp, W., Kummu, M., Mohan, C., Nogués-Bravo, D., ... Rockström, J. (2023). Earth beyond six of nine planetary boundaries. *Science Advances, 9*(37), eadh2458. https://doi.org/10.1126/sciadv.adh2458.

Ridgen, D., & Rossier, N. (Directors). (2009). *American Radical: The Trials of Norman Finkelstein* [Video recording]. https://www.imdb.com/title/tt1475191/.

Riedl, K., Jensen, K., Call, J., & Tomasello, M. (2015). Restorative Justice in Children. *Current Biology, 25*(13), 1731–1735. https://doi.org/10.1016/j.cub.2015.05.014.

Riess, H. (2017). The Science of Empathy. *Journal of Patient Experience, 4*(2), 74–77. https://doi.org/10.1177/2374373517699267.

Rifkin, J. (2010). *The Empathic Civilization: The Race to Global Consciousness in a World in Crisis.* Polity.

Riva, P., Brambilla, M., & Vaes, J. (2016). Bad guys suffer less (social pain): Moral status influences judgements of others' social suffering. *The British Journal of Social Psychology, 55*(1), 88–108. https://doi.org/10.1111/bjso.12114.

Rizzolatti, G., & Craighero, L. (2004). The mirror-neuron system. *Annual Review of Neuroscience, 27*, 169–192. https://doi.org/10.1146/annurev.neuro.27.070203.144230.

Roberts, H. (2024, February 18). *The Zone Of Interest producer: Gaza conflict reminds us of 'selective empathy'.* https://www.standard.co.uk/culture/film/james-wilson-gaza-baftas-british-german-b1139891.html.

Robeznieks, A. (2022, March 22). *Senseless war in Ukraine sparks physician aid response.* American Medical Association. https://www.ama-assn.org/public-health/injury-violence-prevention/senseless-war-ukraine-sparks-physician-aid-response.

Romanello, M., McGushin, A., Di Napoli, C., Drummond, P., Hughes, N., Jamart, L., Kennard, H., Lampard, P., Solano Rodriguez, B., Arnell, N., Ayeb-Karlsson, S., Belesova, K., Cai, W., Campbell-Lendrum, D., Capstick, S., Chambers, J., Chu, L., Ciampi, L., Dalin, C., ... Hamilton, I. (2021). The 2021 report of the Lancet Countdown on health and climate change: Code red for a healthy future. *Lancet (London, England), 398*(10311), 1619–1662. https://doi.org/10.1016/S0140-6736(21)01787-6.

Roosevelt, T. (1889). *The winning of the West (Vol. 3).* G. P. Putnam's Sons. https://books
.google.bs/books?id=vMFYAAAAMAAJ&printsec=frontcover&source=gbs_ge
_summary_r&cad=0#v=onepage&q&f=false.

Russell, B. (1950, September 3). If We Are to Survive This Dark Time—; Bertrand Russsell advises as to learn, to look at things 'under the aspect of eternity. *The New York Times.* https://www.nytimes.com/1950/09/03/archives/if-we-are-to-survive -this-dark-time-bertrand-russsell-advises-as-to.html.

Russell, B., & Einstein, A. (1955). *Statement: The Russell-Einstein Manifesto.* Pugwash Conferences on Science and World Affairs. https://pugwash.org/1955/07/09 /statement-manifesto/.

Sachs, J. (2024). *Comment of 'NATO: What You Need To Know | Medea Benjamin & David Swanson' OR Books.* https://www.counterfire.org/article/nato-what-you-need -to-know-book-review/.

Sadek, D., & Digital Forensic Research Lab. (2024, February 14). Suspicious accounts on X amplify allegations against UNRWA. *DFRLab.* https://dfrlab.org/2024/02/14 /suspicious-accounts-on-x-amplify-allegations-against-unrwa/.

Sah, S. (2024). Infectious diseases are being allowed to run rampant in Gaza. *BMJ, 387,* q2186. https://doi.org/10.1136/bmj.q2186.

Sah, S., & Dawas, K. (2024). *Israel is using starvation as a weapon of war in Gaza.* https:// doi.org/10.1136/bmj.q1018.

Said, E. W. (1992). *The Question of Palestine.* Knopf Doubleday Publishing Group.

Sainato, M. (2024, October 24). US professors face discipline and investigations over Palestine support. *The Guardian.* https://www.theguardian.com/us-news/2024 /oct/24/university-professors-discipline-palestine-support.

Salam, E. (2023, November 10). Outrage grows after 'chilling call for genocide' by Florida Republican. *The Guardian.* https://www.theguardian.com/us-news/2023 /nov/10/florida-republican-michelle-salzman-palestine.

Samudzi, Z. (2024). "We are Fighting Nazis": Genocidal Fashionings of Gaza(ns) After 7 October. *Journal of Genocide Research, 0*(0), 1–9. https://doi.org/10.1080/14623528 .2024.2305524.

Sanderson, H., Czub, M., Jakacki, J., Koschinski, S., Tougaard, J., Sveegaard, S., Frey, T., Fauser, P., Bełdowski, J., Beck, A. J., Przyborska, A., Olejnik, A., Szturomski, B., & Kicinski, R. (2023). Environmental impact of the explosion of the Nord Stream pipelines. *Scientific Reports, 13*(1), 19923. https://doi.org/10.1038/s41598-023-47290-7.

Sapolsky, R. (2019, February 12). *This Is Your Brain on Nationalism.* Foreign Affairs. https://www.foreignaffairs.com/world/your-brain-nationalism.

Sarfraz, H. (2024, June 4). *Genocide in plain sight.* Tribune The Express. https://www .genocidewatch.com/single-post/genocide-in-plain-sight.

Sarid, Y. (2015, January 23). This Israeli politician is basically a psychopath. Guess who. *Haaretz.* https://www.haaretz.com/opinion/2015-01-23/ty-article/.premium /israels-candidates-must-be-crazy/0000017f-e719-df5f-a17f-ffdff6560000.

Save the Children. (2025, January 14). *GAZA: EXPLOSIVE WEAPONS LEFT 15 CHILDREN A DAY WITH POTENTIALLY LIFELONG DISABILITIES IN 2024.* Save the Children International. https://www.savethechildren.net/news/gaza-explosive-weapons-left-15-children-day-potentially-lifelong-disabilities-2024.

Savell, S. (2023, May 15). *How Deaths Outlive War—The Reverberating Impact of Post 9/11 Wars on Human Health* [Watson Institute International & Public Affairs, Brown University]. The Costs of War. https://watson.brown.edu/costsofwar/figures/2023/IndirectDeaths.

Savitsky, Z. (2022, August 5). *Nuclear war would cause yearslong global famine | Science | AAAS*. Science. https://www.science.org/content/article/nuclear-war-would-cause-yearslong-global-famine.

Scahill, J. (2016). *The Assassination Complex: Inside the Government's Secret Drone Warfare Program*. Simon and Schuster.

Seidman, D. (2024, November 23). *Jewish Anti-Zionist Activist Describes His Arrest Under UK's Anti-Terror Law*. Truthout. https://truthout.org/articles/jewish-anti-zionist-activist-describes-his-arrest-under-uks-anti-terror-law/.

Shapiro, Y. A. (2020). *The Empty Wagon: Zionism's Journey from Identity Crisis to Identity Theft*. Bais Medrash Society.

Shaw, M. (2024). Inescapably Genocidal. *Journal of Genocide Research*, 0(0), 1–5. https://doi.org/10.1080/14623528.2023.2300555.

Shehab, E. (2025, April 10). *Do you know what unsettles me most about this war? Not just the bombs. Not just the starvation. Not just the cold, calculated siege of two million souls. It's the silence. It's the choking, complicit silence of a world that once branded the 20th century as the "century of* [Dr. Ezzideen [@ezzingaza]]". Twitter. https://x.com/ezzingaza/status/1910453896293539938.

Sherwood, H. (2014, July 20). *Israelis gather on hillsides to watch and cheer as military drops bombs on Gaza | Israel*. The Guardian. https://www.theguardian.com/world/2014/jul/20/israelis-cheer-gaza-bombing.

Shlaim, A. (1995). *War and Peace in the Middle East: A Concise History, Revised and Updated*. Penguin Publishing Group.

Shlaim, A. (1998). *The Politics of Partition: King Abdullah, the Zionists, and Palestine, 1921–1951*. Oxford University Press.

Shogan, C. J. (2009). The Contemporary Presidency: The Political Utility of Empathy in Presidential Leadership. *Presidential Studies Quarterly*, 39(4), 859–877. https://doi.org/10.1111/j.1741-5705.2009.03711.x.

Shulman, D. (2015, November 4). Jerusalem: Why Should Things Not Get Worse? *The New York Review of Books*. https://www.nybooks.com/online/2015/11/04/jerusalem-why-should-things-not-get-worse/.

Siddique, H. (2025, May 10). *UK Lawyers for Israel condemned over claim war may reduce obesity in Gaza | Israel-Gaza war | The Guardian*. https://www.theguardian.com/world/2025/may/10/uk-lawyers-for-israel-condemned-over-claim-war-may-reduce-obesity-in-gaza.

Sidhwa, F. (2024, October 9). Opinion | 65 Doctors, Nurses and Paramedics: What We Saw in Gaza. *The New York Times*. https://www.nytimes.com/interactive/2024/10/09/opinion/gaza-doctor-interviews.html.

Skeem, J. L., Polaschek, D. L. L., Patrick, C. J., & Lilienfeld, S. O. (2011). Psychopathic Personality: Bridging the Gap Between Scientific Evidence and Public Policy. *Psychological Science in the Public Interest: A Journal of the American Psychological Society*, 12(3), 95–162. https://doi.org/10.1177/1529100611426706.

Sky News. (2025, January 20). *Three Israeli hostages returned and 90 Palestinian detainees released as part of ceasefire deal | World News | Sky News*. https://news.sky.com/story/british-israeli-woman-among-first-three-hostages-hamas-plans-to-release-today-13292111.

Small, D. A., & Loewenstein, G. (2003). Helping a Victim or Helping the Victim: Altruism and Identifiability. *Journal of Risk and Uncertainty*, 26(1), 5–16. https://doi.org/10.1023/A:1022299422219.

Smerkovich, L. S. and M. (2024, May 30). Israeli Views of the Israel-Hamas War. *Pew Research Center*. https://www.pewresearch.org/global/2024/05/30/israeli-views-of-the-israel-hamas-war/.

Smith A. (1776). *An Inquiry into the Nature and Causes of the Wealth of Nations Vol II*. Printed for W. Strahan; and T. Cadell.

Sondos, A. (2025, June 5). *Israel committing genocide in Gaza, says top legal scholar Melanie O'Brien*. Middle East Eye. https://www.middleeasteye.net/news/israel-committing-genocide-gaza-says-top-legal-scholar-melanie-obrien.

Spencer, H. (1995). *Social statics: The conditions essential to human happiness specified, and the first of them developed*. Robert Schalkenbach Foundation.

Stannard, D. E. (1992). *American Holocaust: The Conquest of the New World*. Oxford University Press, USA.

Starblanket, T. (2018). *Suffer the Little Children: Genocide, Indigenous Nations and the Canadian State*. Clarity Press, Incorporated.

Stavrou, A. (2024, November 1). *More than 100 BBC staff accuse broadcaster of Israel bias in Gaza coverage*. The Independent. https://www.independent.co.uk/news/uk/home-news/bbc-israel-gaza-letter-tim-davie-bias-palestine-b2636737.html.

Stellar, J., Feinberg, M., & Keltner, D. (2014). When the selfish suffer: Evidence for selective prosocial emotional and physiological responses to suffering egoists. *Evolution and Human Behavior*, 35(2), 140–147. https://doi.org/10.1016/j.evolhumbehav.2013.12.001.

Stuart, R. (2024, October 8). *'I'm so scared, please come': Heartbreaking final moments of girl, 5, killed in Gaza*. Sky News. https://news.sky.com/story/im-so-scared-please-come-heartbreaking-final-moments-of-girl-5-killed-in-gaza-13229813.

Subramanian, S. V., Jones, K., Kaddour, A., & Krieger, N. (2009). Revisiting Robinson: The perils of individualistic and ecologic fallacy. *International Journal of Epidemiology*, 38(2), 342–360. https://doi.org/10.1093/ije/dyn359.

Sullivan, D., & Hickel, J. (2023). Capitalism and extreme poverty: A global analysis of real wages, human height, and mortality since the long 16th century. *World Development, 161*, 106026. https://doi.org/10.1016/j.worlddev.2022.106026.

Sullivan F. (1995). *Indian Freedom: The Cause of Bartolome de Las Casas: A Reader.* Rowman & Littlefield.

Sultany, N. (2024). A Threshold Crossed: On Genocidal Intent and the Duty to Prevent Genocide in Palestine. *Journal of Genocide Research,* 1–26. https://doi.org/10.1080/14623528.2024.2351261.

Surian, L., Parise, E., & Geraci, A. (2025). Core Moral Concepts and the Sense of Fairness in Human Infants. *Human Nature (Hawthorne, N.Y.), 36*(1), 121–142. https://doi.org/10.1007/s12110-025-09490-0.

Svetlova, K. (2024, May 21). The Empathy Gap. *The Jerusalem Strategic Tribune.* https://jstribune.com/svetlova-the-empathy-gap/.

Tayler, L. (2013). Between a Drone and Al-Qaeda. *Human Rights Watch.* https://www.hrw.org/report/2013/10/22/between-drone-and-al-qaeda/civilian-cost-us-targeted-killings-yemen.

Tercatin, R. (2024, November 8). *Violenze ad Amsterdam, Netanyahu: "È tornata la notte dei cristalli". Israele: "Un attacco premeditato, avevamo avvertito l'Olanda".* La Repubblica. https://www.repubblica.it/esteri/2024/11/08/news/scontri_amsterdam_israele_attaccoolanda-423605140/.

Terman, L. M. (1916). *The Measurement of Intelligence: An Explanation of and a Complete Guide for the Use of the Stanford Revision and Extension of the Binet-Simon Intelligence Scale.* Houghton Mifflin.

Tesfaye, S. (2017, September 19). *New study: White people lack empathy across the socioeconomic spectrum.* Salon. https://www.salon.com/2017/09/19/new-study-white-people-lack-empathy-across-the-socioeconomic-spectrum/.

The Economist. (2022, April 13). What is genocide, and is Russia committing it in Ukraine? *The Economist.* https://www.economist.com/the-economist-explains/2022/04/13/what-is-genocide-and-is-russia-committing-it-in-ukraine?utm_medium=cpc.adword.pd&utm_source=google&ppccampaignID=18151738051&ppcadID=&utm_campaign=a.22brand_pmax&utm_content=conversion.direct-response.anonymous&gad_source=1&gclid=CjwKCAjwjeuyBhBuEiwAJ3vuoVcwl8CzPY4rRgjjUbuma_AY8b7GufynLAahb1Df1niWVmJdIS5DbxoCuBoQAvD_BwE&gclsrc=aw.ds.

The Economist. (2023, October 18). *What is Palestinian Islamic Jihad? Israel blames the group for a deadly explosion at a hospital in Gaza.* https://www.economist.com/the-economist-explains/2023/10/18/what-is-palestinian-islamic-jihad.

The Economist. (2024, January 18). *Charging Israel with genocide makes a mockery of the ICJ.* https://www.economist.com/leaders/2024/01/18/charging-israel-with-genocide-makes-a-mockery-of-the-icj.

The Economist. (2025, May 8). *The war in Gaza must end.* https://www.economist.com/leaders/2025/05/08/the-war-in-gaza-must-end.

The Financial Times. (2025, May 8). *The west's shameful silence on Gaza.* Archive.Is. https://archive.is/X6Rtz.

The Grayzone (Director). (2024, March 21). *'Kill them all': Inside the Israeli blockade on Gaza aid* [Video recording]. https://www.youtube.com/watch?v=LqRzfb2oMaM.

The Hill. (2024, June 4). *'Israelis look like us': Bill Maher wields identity politics to defend Israel: Briahna Joy Gray.* The Hill. https://thehill.com/video/%E2%80%98israelis-look-like-us%E2%80%99-bill-maher-wields-identity-politics-to-defend-israel-briahna-joy-gray/9754183/.

The Jerusalem Post. (2025, February 3). *Approx. 80% of Israelis support Trump's plan to relocate Gazans—Survey.* The Jerusalem Post. https://www.jpost.com/international/article-840500.

The Journal, P. (2025, April 15). *'We are complicit': Irish WHO director calls for action on war in Gaza.* TheJournal.Ie. https://www.thejournal.ie/irish-who-director-calls-for-action-war-in-gaza-6695092-May2025/.

Theroux, L. (2025). *Louis Theroux: The Settlers.* https://www.imdb.com/it/title/tt36640622/.

Times of Israel. (2025, February 25). *Bibas family tells Netanyahu to 'shut up,' as he details the murders of Shiri, Ariel and Kfir.* https://www.timesofisrael.com/bibas-family-tells-netanyahu-to-shut-up-as-he-details-the-murders-of-shiri-ariel-and-kfir/.

Tkacik, M. (2024, March 20). *What Really Happened on October 7?* The American Prospect. https://prospect.org/api/content/604afa20-e6fe-11ee-90f2-12163087a831/.

Trent, N. L., Beauregard, M., & Schwartz, G. E. (2020). Preliminary development and validation of a scale to measure universal love. *Spirituality in Clinical Practice, 7*(1), 51–64. https://doi.org/10.1037/scp0000198.

Turse, N. (2025, April 1). *News Graveyards: How Dangers to War Reporters Endanger the World | Costs of War.* The Costs of War. https://watson.brown.edu/costsofwar/papers/2025/Journalists.

TWAILR. (2023, October 17). *Public Statement: Scholars Warn of Potential Genocide in Gaza.* Third World Approaches to International Law Review. https://twailr.com/public-statement-scholars-warn-of-potential-genocide-in-gaza/.

UN. (2023, October 24). *Secretary-General's remarks to the Security Council—On the Middle East [as delivered] | United Nations Secretary-General.* https://www.un.org/sg/en/content/sg/statement/2023-10-24/secretary-generals-remarks-the-security-council-the-middle-east-delivered.

UN Human Rights Council. (2009). *Report of the United Nations Fact Finding Mission on the Gaza Conflict (Advance Edited Version).*

UN Human Rights Council. (2024, July 1). Anatomy of a Genocide—Report of the Special Rapporteur on the situation of human rights in the Palestinian territory occupied since 1967 to Human Rights Council. *Question of Palestine.* https://docs .un.org/en/A/HRC/55/73.

UNICEF. (2024, October 18). *Gaza's children: 'Trapped in a cycle of pain'.* https://www .unicef.org/press-releases/gazas-children-trapped-cycle-pain.

UNICEF. (2025, May 2). *Statement from UNICEF Executive Director Catherine Russell on the situation for children in the Gaza Strip after two months of aid blockade.* https:// www.unicef.org/press-releases/statement-unicef-executive-director-catherine -russell-situation-children-gaza-strip.

United Nations. (2024a). *Mission report: Official visit of the Office of the SRSG-SVC to Israel and the occupied West Bank : 29 January–14 February 2024* (UN Special Representative of the Secretary-General on Sexual Violence In). UN,. https:/ /digitallibrary.un.org/record/4039262.

United Nations. (2024b, April 22). *"A war on the right to health"* [United Nations—The Question of Palestine]. https://www.un.org/unispal/document/un-expert-raised -concern-about-gazans-mental-health-due-to-conflict/.

United Nations. (2024c, April 25). *Amid campus crackdowns, Gaza war triggers freedom of expression crisis.* UN News Global Perspective Human Stories. https://news .un.org/en/story/2024/04/1149001.

United Nations Relief and Works Agency for Palestine Refugees in the Near East. (2025, July 11). *UNRWA Situation Report #179 on the humanitarian crisis in the Gaza Strip and the West Bank, including East Jerusalem.* https://www.unrwa.org/resources/ reports/unrwa-situation-report-179-situation-gaza-strip-and-west-bank-including -east-jerusalem.

UNOCHA. (2014, August 28). *Occupied Palestinian Territory: Gaza Emergency Situation Report (as of 28 August 2014, 08:00 hrs).* United Nations Office for the Coordination of Humanitarian Affairs—Occupied Palestinian Territory. https://www .ochaopt.org/content/occupied-palestinian-territory-gaza-emergency-situation -report-28-august-2014-0800-hrs.

UNOCHA. (2024a). *Six-month update report on the human rights situation in Gaza: 1 November 2023 to 30 April 2024.* https://www.ohchr.org/en/documents /reports/six-month-update-report-human-rights-situation-gaza-1-november -2023-30-april-2024.

UNOCHA. (2024b, April 19). *Hostilities in the Gaza Strip and Israel—Reported humanitarian impact, 19 April 2024 at 15:00 | OCHA.* https://www.unocha.org /publications/report/occupied-palestinian-territory/hostilities-gaza-strip-and -israel-reported-humanitarian-impact-19-april-2024-1500.

UNOCHA. (2024c, October 9). *'We cannot afford to look away': OCHA urges Security Council to end violence in Gaza | OCHA.* https://www.unocha.org/news /we-cannot-afford-look-away-ocha-urges-security-council-end-violence-gaza.

UNOCHA. (2025). *Reported impact snapshot | Gaza Strip (25 March 2025)* (United Nations Office for the Coordination of Humanitarian Affairs). https://www.ochaopt.org/content/reported-impact-snapshot-gaza-strip-25-march-2025.

UNOCHA. (2025a). *Data on casualties | United Nations Office for the Coordination of Humanitarian Affairs—Occupied Palestinian Territory.* https://www.ochaopt.org/data/casualties.

UNOCHA. (2025b, January 28). *Humanitarian Situation Update #259 | Gaza Strip.* United Nations Office for the Coordination of Humanitarian Affairs—Occupied Palestinian Territory. https://www.ochaopt.org/content/humanitarian-situation-update-259-gaza-strip.

UNOCHA. (2025c, March 25). *Humanitarian Situation Update #275 | Gaza Strip.* United Nations Office for the Coordination of Humanitarian Affairs—Occupied Palestinian Territory. https://www.ochaopt.org/content/humanitarian-situation-update-275-gaza-strip.

UNOSAT. (2025, October). *UNOSAT Gaza Strip Comprehensive Damage Assessment.* https://unosat.org/products/4205.

Vachon, D. D., & Lynam, D. R. (2016). Fixing the Problem With Empathy: Development and Validation of the Affective and Cognitive Measure of Empathy. *Assessment, 23*(2), 135–149. https://doi.org/10.1177/1073191114567941.

Vachon, D. D., Lynam, D. R., & Johnson, J. A. (2014). The (non)relation between empathy and aggression: Surprising results from a meta-analysis. *Psychological Bulletin, 140*(3), 751–773. https://doi.org/10.1037/a0035236.

Waal, F. de. (2009). *The Age of Empathy: Nature's Lessons for a Kinder Society.* Crown.

Wagner, L. (2024, April 29). *Journalism professors call on New York Times to review Oct. 7 report.* The Washington Post. https://www.washingtonpost.com/style/media/2024/04/29/new-york-times-oct-7-journalism-professors-letter/.

Walker, J. (2024, November 8). *White House denies Dems 'abandoned' working class after scathing Bernie Sanders statement.* WBMA. https://abc3340.com/news/nation-world/white-house-denies-dems-abandoned-working-class-after-scathing-bernie-sanders-statement-presidential-election-kamala-trump-biden-democratic-party-karine-jean-pierre.

Wall Street Journal. (2023a, October 14). *Airstrike Hits Convoy of Gaza Civilians Fleeing South.* WSJ. https://www.wsj.com/video/airstrike-hits-convoy-of-gaza-civilians-fleeing-south/3F735CB9-1162-4304-A1B0-FF4232A38AD7?fbclid=IwAR0Jpz14QXIyPIhQTgD6qGG1XUSa6xBtNk8eB-fY-U58K2lYclVypi4La1g_aem_AeDRHWc2MbM1zOCX3Hz1KsfGymVQ2XypQlslcYFdLuHWWYoOAHvIWjGGPINHfqSfuDs.

Wall Street Journal. (2023b, October 22). *Video Analysis Shows Gaza Hospital Hit By Failed Rocket Meant for Israel.* https://www.wsj.com/livecoverage/israel-hamas-war-gaza-strip-conflict/card/watch-video-analysis-shows-gaza-hospital-was-hit-by-failed-rocket-meant-for-israel-rP4uhNqD5MQoYryXxsvC.

Wang, A., & Todd, A. (2021). Evaluations of empathizers depend on the target of empathy. *Journal of Personality and Social Psychology, 121*(5), 1005–1028. https://doi.org/10.1037/pspi0000341.

Wang, Y., Harris, P. L., Pei, M., & Su, Y. (2022). Do Bad People Deserve Empathy? Selective Empathy Based on Targets' Moral Characteristics. *Affective Science, 4*(2), 413–428. https://doi.org/10.1007/s42761-022-00165-y.

Wang, Y., Harris, P. L., Pei, M., & Su, Y. (2023). Do Bad People Deserve Empathy? Selective Empathy Based on Targets' Moral Characteristics. *Affective Science, 4*(2), 413–428. https://doi.org/10.1007/s42761-022-00165-y.

Wang, Y., Zheng, D., Chen, J., Rao, L.-L., Li, S., & Zhou, Y. (2019). Born for fairness: Evidence of genetic contribution to a neural basis of fairness intuition. *Social Cognitive and Affective Neuroscience, 14*(5), 539–548. https://doi.org/10.1093/scan/nsz031.

War Child. (2024, December 11). *War Child shares first study of psychological impact of war on vulnerable children in Gaza—News—War Child.* https://www.warchild.org.uk/news/war-child-shares-first-study-psychological-impact-war-vulnerable-children-gaza.

Warikoo, N. (2024). *Trump wins Dearborn, makes gains in Hamtramck amid anger over Gaza.* Detroit Free Press. https://eu.freep.com/story/news/politics/elections/2024/11/06/trump-wins-dearborn-and-makes-gains-in-hamtramck/76085841007/.

Warnke, K., Martinović, B., & Rosler, N. (2024). Territorial ownership perceptions and reconciliation in the Israeli–Palestinian conflict: A person-centred approach. *European Journal of Social Psychology, 54*(1), 31–47. https://doi.org/10.1002/ejsp.2993.

Warschawski, M. (2004, December 1). Monthly Review | The New Israel. *Monthly Review.* https://monthlyreview.org/2004/12/01/the-new-israel/.

Watson, L. (2022, March 3). *The lifesaving strangers offering refuge to Ukrainians fleeing the war.* ITV News. https://www.itv.com/news/2022-03-03/the-lifesaving-strangers-offering-refuge-to-ukrainians-fleeing-the-war.

Watts, A., & Lilienfeld, S. O. (2016, January 26). *Not all psychopaths are criminals—some psychopathic traits are actually linked to success.* The Conversation. http://theconversation.com/not-all-psychopaths-are-criminals-some-psychopathic-traits-are-actually-linked-to-success-51282.

WHO. (2022). *Right to Health Barriers to health and attacks on health care in the occupied Palestinian territory, 2019 to 2021* (World Health Organization). https://applications.emro.who.int/docs/9789292740887-eng.pdf?ua=1.

WHO. (2023, October 14). *Evacuation orders by Israel to hospitals in northern Gaza are a death sentence for the sick and injured.* World Health Organization—Regional Office for the Eastern Mediterranean. http://www.emro.who.int/media/news/evacuation-orders-by-israel-to-hospitals-in-northern-gaza-are-a-death-sentence-for-the-sick-and-injured.html.

WHO. (2024, December 28). *Kamal Adwan Hospital out of service following a raid yesterday and repeated attacks since October.* World Health Organization. https://hq_who_departmentofcommunications.cmail19.com/t/d-e-syddtld-jdyhydttc-y/.

WHO. (2025, May 22). *Gaza Hostilities 2023 / 2024—Emergency Situation Reports.* World Health Organization—Regional Office for the Eastern Mediterranean. https://www.emro.who.int/images/stories/Sitrep_59.pdf?ua=1.

Wilde, O. (with Project Gutenberg). (1892). *Lady Windermere's Fan.* http://archive.org/details/ladywindermeresfoo79ogut.

Willcock, S., Cooper, G. S., Addy, J., & Dearing, J. A. (2023). Earlier collapse of Anthropocene ecosystems driven by multiple faster and noisier drivers. *Nature Sustainability,* 6(11), Article 11. https://doi.org/10.1038/s41893-023-01157-x.

Wille, B. (2024). "I Can't Erase All the Blood from My Mind". *Human Rights Watch.* https://www.hrw.org/report/2024/07/17/i-cant-erase-all-blood-my-mind/palestinian-armed-groups-october-7-assault-israel.

Winstanley, A. (2023). *Weaponising Anti-Semitism: How the Israel Lobby Brought Down Jeremy Corbyn.* OR Books.

Wintour, P. (2025, May 13). No evidence of genocide in Gaza, UK lawyers say in arms export case. *The Guardian.* https://www.theguardian.com/world/2025/may/13/no-evidence-of-genocide-in-gaza-uk-lawyers-say-in-arms-export-case.

Wispelwey, B., Osuagwu, C., Mills, D., Goronga, T., & Morse, M. (2023). Towards a bidirectional decoloniality in academic global health: Insights from settler colonialism and racial capitalism. *The Lancet Global Health,* 11(9), e1469–e1474. https://doi.org/10.1016/S2214-109X(23)00307-8.

Wispelwey, B., Tanous, O., Asi, Y., Hammoudeh, W., & Mills, D. (2023). Because its power remains naturalized: Introducing the settler colonial determinants of health. *Frontiers in Public Health,* 11. https://doi.org/10.3389/fpubh.2023.1137428.

Witus, Z. (2024, October 17). *Sha'ban al-Dalou burned alive before the world. May his death awaken us | Zak Witus | The Guardian.* https://www.theguardian.com/commentisfree/2024/oct/17/shaban-al-dalou-burned-alive-israel-gaza.

Witze, A. (2020). How a small nuclear war would transform the entire planet. *Nature,* 579(7800), 485–487. https://doi.org/10.1038/d41586-020-00794-y.

Wootliff, R. (2018, July 19). *Israel passes Jewish state law, enshrining 'national home of the Jewish people'.* https://www.timesofisrael.com/knesset-votes-contentious-jewish-nation-state-bill-into-law/.

World Economic Forum. (2023, January 11). *Global Risks Report 2023.* https://www.weforum.org/publications/global-risks-report-2023/.

Worthington, A. (2024, October 25). *Gaza: The Normalization of Genocide Through the Complete Complicity of the West and Its Colonization by Israel | Andy Worthington.* https://www.andyworthington.co.uk/2024/10/25/gaza-the-normalization-of-genocide-through-the-complete-complicity-of-the-west-and-its-colonization-by-israel/.

Writers Against the War on Gaza. (2024a, August 16). *'Words like Slaughter:' A comparative study of The New York Times reporting in Ukraine and Gaza.* Mondoweiss.

https://mondoweiss.net/2024/08/words-like-slaughter-a-comparative-study-of
-the-new-york-times-reporting-in-ukraine-and-gaza/.

Writers Against the War on Gaza. (2024b, August 16). *"Words Like Slaughter": A Study of The New York Times' Reporting in Ukraine and Gaza.* The New York War Crimes | "All the Consent That's Fit to Manufacture". https://newyorkwarcrimes.com/words -like-slaughter.

Xia, L., Robock, A., Scherrer, K., Harrison, C. S., Bodirsky, B. L., Weindl, I., Jägermeyr, J., Bardeen, C. G., Toon, O. B., & Heneghan, R. (2022). Global food insecurity and famine from reduced crop, marine fishery and livestock production due to climate disruption from nuclear war soot injection. *Nature Food, 3*(8), 586–596. https:// doi.org/10.1038/s43016-022-00573-0.

Xu, Y., & Ramanathan, V. (2017). Well below 2 °C: Mitigation strategies for avoiding dangerous to catastrophic climate changes. *Proceedings of the National Academy of Sciences, 114*(39), 10315–10323. https://doi.org/10.1073/pnas.1618481114.

Yang, Y., Zhan, J., & Fan, Y. (2025). From national identity to well-being: The crucial mediating role of self-esteem in adolescents. *BMC Psychology, 13*(1), 507. https:// doi.org/10.1186/s40359-025-02776-z.

Yehuda Shaul [@YehudaShaul]. (2023, October 25). *[3] "there should be 2 goals for this victory: 1. There is no more Muslim land in the Land of Israel ... After we make it the land of IL, Gaza should be left as a monument, like Sodom ..." MK Amit Halevi, Likud* https://t.co/q8hk4W7i2r [Tweet]. Twitter. https://twitter.com/YehudaShaul /status/1717219175054115240.

Yiftachel, O. (2006). *Ethnocracy: Land and Identity Politics in Israel/Palestine.* University of Pennsylvania Press.

Zaki, J. (2014). Empathy: A motivated account. *Psychological Bulletin, 140*(6), 1608–1647. https://doi.org/10.1037/a0037679.

Zhang, Y., Yu, J., Zhang, Y., Zhang, Y., Sun, F., Yao, Y., Bai, Z., Sun, H., Zhao, Q., & Li, X. (2023). Emotional Contagion and Social Support in Pigs with the Negative Stimulus. *Animals: An Open Access Journal from MDPI, 13*(20), 3160. https://doi.org/10.3390 /ani13203160.

Zhang, Z., & Bayly, B. (2024). Patterns of early childhood adversity and neighborhood deprivation predict unique challenges in adolescence: A UK birth cohort study. *Development and Psychopathology,* 1–13. https://doi.org/10.1017/S0954579424001561.

Zilber, N. (2024, May 26). Dozens killed and wounded after explosions at Gaza 'safe-zone' camp. *Financial Times.*

Zinn H. (1980). *A People's History of the United States.* Harper Perennial—Modern Classics.

Zipfel, M., & Nehme, A. (2024, February 8). Watchdog org. Claims NYU violated policy in faculty suspension. *Washington Square News.* https://nyunews.com/news /2024/02/08/fire-letter-israel-hamas-conflict/.

Ziv, O. (2025, March 24). *'When our grandchildren ask about the genocide, I'll say I refused'.* +972 Magazine. https://www.972mag.com/ella-keidar-greenberg-israeli -military-refusal/.